FOREST AESTHETICS

Heinrich von Salisch

Translated by
Walter L. Cook Jr. and Doris Wehlau

From the second enlarged edition
with 16 photographs and
numerous pictures in the text.
Publisher Julius Springer, Berlin, 1902

The Forest History Society is a nonprofit educational and research institution dedicated to the advancement of historical understanding of human interaction with the forest environment. It was established in 1946. Interpretations and conclusions in FHS publications are those of the authors; the institution takes responsibility for the selection of topics, competency of the authors, and their freedom of inquiry.

This book was published with support from the Lynn W. Day Endowment for Forest History Publications, Edward W. Phares, and Walter L. Cook.

Printed in the United States of America
Forest History Society
701 William Vickers Avenue
Durham, North Carolina 27701
(919) 682-9319
www.foresthistory.org

First edition, 1885
Second enlarged edition, 1902
Third edition, 1911

Library of Congress Cataloging-in-Publication Data

Salisch, Heinrich von.
[Forstästhetik. English]
Forest aesthetics / Heinrich von Salisch ; translated by Walter L. Cook Jr. and Doris Wehlau.
p. cm.
"From the second enlarged edition with 16 photographs and numerous pictures in the text. Publisher Julius Springer, Berlin, 1902."
Includes bibliographical references and index.
Summary: "This book is a translation of the 1902 German text Forstästhetik by von Salisch. It represents the earliest thinking about forest aesthetics. It provides a window to the origins of the landscape architecture and forestry professions and a historical context for those seeking both beauty and income from the forest" --Provided by the publisher.
ISBN 978-0-89030-072-5 (pbk. : alk. paper)
1. Forests and forestry. 2. Nature (Aesthetics) I. Forest History Society. II. Title.
SB476.S313 2008
634.9--dc22

2008028536

Motto:

For the minor head it always will be a craft,
for the better one an art.

Goethe, Wilhelm Meisters Wanderjahre

ACKNOWLEDGEMENTS

Doris Wehlau and I received much critical help with the work of translating *Forest Aesthetics.* Most important was the permission of my dean, Leon Hargreaves, to spend my time on the project without funding as a research project. Similarly, Doris's time was considered as academic study, and she received ten quarter hours of credit. We used the computers and space of the School of Forest Resources at the University of Georgia. Also critical was the availability of the original German texts through inter-library loan; these were photocopied using the School's copier and paper. The university's main library generously allowed us the long-term loan of Fluegel's *Woerterbuch*, 4th edition, an 1894 German-to-English dictionary. Two other reference works were also essential to the success of the venture: Dr. Klaus Steinbeck's German-to-English 1938 glossary of forestry terms and *Krussman's Manual of Cultivated Broad-leaf Trees and Plant Communities of the Southeast.*

The foregoing items pertained directly to the translation. Several other sources of assistance were critical to the study and tour of the von Salisch estate and social environment. First, I discovered, with the help of the university's map room librarian, a series of finely detailed maps of the inter-war border area between Germany and Poland. These showed the Kreistadt Militsch, the hamlet of Postel, and the palace and surrounding park that was the von Salisch home, as well as the author's forest holdings and the adjacent federal forest. The *Jagen*, or forest compartments, were delineated, numbered, and symbolized with the appropriate species.

The second stroke of luck was the internship of Linda Robertson, Doris's classmate, with the Polish Board for the Preservation of Historic Gardens and Palaces, headquartered in Warsaw. Linda, her supervisor, Tomasz Zweich, and his co-worker, Jakub Zemla, gave me a guided (and translated) tour of Prince Pueckler's estate on the Neisse River, then introduced Linda and me to Ewald Ranoscek and Marek Grobelny, foresters in Militsch. They managed the former von Salisch estate and forests and gave us a tour of the area. Linda had previously sent a packet of maps and other literature about Postel, which was translated for us by Irina Popiashvili from the Republic of Georgia and a graduate student at the University of Georgia. Jacek Siry, a recent member of the Forest Resources faculty, translated letters and some photographed signs from Militsch and Postel.

The long interval between the completion of the translation and writing in 1994, and the publication date is due to many false starts and delays. A truly serendipitous event led me to know Anne Boston, a book designer for a local publisher. She eagerly volunteered to re-type the entire book and get it into condition for publication. When the University of Georgia declined the book for the second time, we were resigned to self-publication. Then I became aware of the Forest History Society. Steven Anderson, president, has worked diligently to find funding, and we owe the final result to him.

Walter L. Cook, Jr.

Contents

Preface for the First Edition ix
Preface for the Second Edition xi
Notes on the Translation xiii
Foreword: The More Things Change . . . xix

PART ONE: Fundamental Principles of Forest Aesthetics

SECTION A. Preliminary Terms

Chapter One. Term and Task of Forest Aesthetics as a Special Forest Science; the Necessity to Study it 1

1. What to Understand by Forest Aesthetics 1
2. It is Necessary to Consider Aesthetic Points of View Everywhere in Forest Management 4
3. History and Literature of Forest Aesthetics. Necessity to Teach and to Study Forest Aesthetics as a Special Branch of Forestry Sciences 12

Chapter Two. The Causes of Delight in Beauty 15

1. Man's Intrinsic Ability to Get Pleasure from the Beautiful 15
2. The Properties Which are Attached to Things and Make up Their Beauty; Teaching of Adornment 19
3. The Pleasure in Beauty Can be Increased in Various Ways; It Needs to be Protected Against Disturbance 30

SECTION B. The Beauty of Nature

Chapter One. Preliminary Remarks on the Relation Between Natural Beauty and Artistic Beauty 34

Chapter Two. Theory of Color in the Landscape 38

Chapter Three. Stones as Adornments of the Forest 54

Chapter Four. The Aesthetic Value of the Tree Species 63

1. Preliminary Remarks 63
2. The Hard Hardwoods 66
3. The Conifers 82
4. The Soft Hardwoods 99
5. The Shrubs 105
6. Foreign Tree Species 110
Chapter Five. Fragrances and Voices of the Forest 113
Final Remarks for PART ONE 117

PART TWO: Applied Forest Aesthetics
SECTION A. Forest Establishment and Forest Economics
Chapter One. Determination of the Most Useful Type of Soil Use 119
Chapter Two. Design of the Road System; Establishing and Naming of Administrative Units 131
Chapter Three. The Silvicultural Systems 149
Chapter Four. Choice of Tree Species 164
Chapter Five. Determination of the Rotation Period 170
Chapter Six. The Regeneration 179
Chapter Seven. Taking Care of the Stand 187
Chapter Eight. The Secondary Utilizations 196
Chapter Nine. Meadows, Water and Fields; Forest Edges, Hedges and Fences 202
SECTION B. Embellishment of the Forest by Special Actions Pursued Particularly in the Interest of Beauty
Chapter One. Park or Forest 213
Chapter Two. The Enhanced Forest 220
Chapter Three. The Establishment and Management of Park-like Landscape 225
Chapter Four. Forest Enhancement by Establishment and Adornment of Roads (Crossroads, Road Signs) 236
Chapter Five. Planting of Trees along Roads and Firebreaks 245
Chapter Six. Old Trees as Adornments of Forests 258

Chapter Seven. Aesthetic Use of Foreign Tree Species and of Varieties of Native Tree Species 267
Chapter Eight. Embellishment of the Forest Stands by Care of the Shrubs and the Groundcover 279
Chapter Nine. Embellishment of the Forests by Stone Blocks 284
Chapter Ten. Monuments, Ruins, Entrenchments 290
Chapter Eleven. Vistas 291
Appendices 301
A. Plates 301
B. Figures and Pictures 303
C. References, 2nd Edition, 1902 305
D. References, 1st Edition, 1885 322
E. Review of 3rd Edition, 1911 343
Index 347

Preface for the First Edition

Whoever might be suited to the great task of writing on forest aesthetics should have achieved a rich knowledge of forestry, he should know well about the philosophical teaching about beauty, and with passion and engagement for the matter, he should have had enough opportunities and time to practice his own ideas in a large area and to test their correctness by the success of the experiments. And, after all, he should have time and a liking to write, which is not everybody's favor.

Considering myself, I quit the administrative career before the head forester examination and have lived since then in the country, far away from any academic life, so that I forgot more than I learned new. My own experiences are limited, for it is less than ten years now that I started to realize methodical forest aesthetic experiments on less than 1,000 ha. Therefore it is obvious that I cannot regard myself as being suited well enough for the task which I set myself out of passion for the matter.

Nevertheless, this did not prohibit me from publishing my writing, for although my book may have many shortcomings and gaps, it is also the only one on this topic.

I have stated this once and for all in the beginning and I intend therefore in the following not to interrupt the context by phrases of modesty which I ought to use (like "according to my humble opinion" or, "if I am allowed to dare to object Mr.NN's opinion, he being a very acknowledged authority. . .").

Only a few things of what I am going to discuss will be new,* but many of the thoughts used here will be found already in the writings listed in the Sources at the end of the book, and some are already practiced every here and there. Nevertheless, I hope that most of it will seem to be new to the dear reader; anyway, I claim for myself the modest merit by having collected this widespread material to have established a serious beginning and to have started the formation of forest aesthetics as an independent branch of forestry sciences, according to my abilities.

Postel near Militsch, January 1885.
v. Salisch.

*The suggestions for thinning out on page 189 will probably contain new points of view.

Preface for the Second Edition

Fifteen years after publication of the first edition, which has been out of print for quite a while, *Forest Aesthetics* is published again in a totally revised and much enlarged shape. If there are considerable gaps even now and some important areas are covered only slightly, I hope this may be kindly excused, for parliamentary duties have kept me very busy since 1893, besides other obligations, so that I have been prohibited from my forestry hobbies completely for some time.

I hope it will not be long until forest aesthetics will be taught at the forestry schools and this field of knowledge will be taken care of and promoted by professionals! In the meantime, I have put special attention on the literature sources in order to facilitate study for other people. A relevant book: Rudorff, Ernst: *Local Preservation,* Leipzig and Berlin, 1901, publisher Georg Heinrich Meyer; I mention it here with regrets because I learned about this treasure trove of appropriate ideas only after I had finished my work.

The same is true for Sohnrey: *Guide for Rural Welfare Work and Local Maintenance,* Berlin, 1901. I learned, too, about this excellent book, which discusses the social relevance of the forests in detail, only after the manuscript had been sent away for printing.

For foresters who do not have the time or passion to read *Forest Aesthetics* in its entirety, but nevertheless would like to hear occasionally about the forest aesthetic judgement of a certain problem, there is an index put together with the utmost possible care. Hopefully, this book will be picked up often and not in vain!

Due to the publisher's kindness, this book has been richly supplied with numerous pictures out of the ateliers of Meisenbach,

Riffarth & Co. (Berlin-Schoneberg), which not only decorate the book but also contribute much to its clarification.

Postel, January 1902.
v. Salisch.

Notes on the Translation

This translation has roots in two centuries, on two continents, and in three persons. The author, Heinrich von Salisch, was a forester and nobleman of rural Silesia, a part of Prussia and the German Empire during his lifetime (1846–1920), and a part of Poland since 1945. As one collaborator of the translation, I am a forester and professor in the Daniel B. Warnell School of Forest Resources at the University of Georgia in Athens. Doris Wehlau is the other collaborator and the primary translator from the original in old-style German typeface; she is a German landscape architect who, during the time of the translation (1992–1993), was a graduate student in the School of Environmental Design at the University of Georgia.

I first learned of *Forest Aesthetics* while doing research for my Master's thesis at the State University of New York College of Forestry at Syracuse University. The book was in the library, and, although I could not read it, I recognized its importance to my thesis topic, *Forest Aesthetics and Timber Harvesting: A Critical Review of the Literature* (1969). Not until 23 years later, when Doris Wehlau enrolled in my course in wilderness management, did the opportunity to discover its contents present itself. Doris was at Georgia studying under a Fulbright Scholarship, and readily accepted the challenge when I told her of the book.

My frustration of not being able to read German was shared by von Salisch, as he regretted not being able to read English. His many references to William Gilpin's 1791 book, *Forest Scenery,* were due to its having been translated into German, and was the only English book on the topic that was available to him. Similarly, since reading *Forest Aesthetics,* I became aware of Prince Hermann Pueckler's book *Hints on Landscape Gardening* (1834), and discovered it had been translated into English and was, in fact, in the University of Georgia library.

The translation of *Forest Aesthetics* will add to the sources available to scholars of forest landscapes of the past. But I use the word past with some hesitation, because in studying *Forest Aesthetics,* I was frequently surprised at the similarity of the thoughts, taste, and problems of late 19th-century Silesia (and broadly, Central Europe) with those of late 20th-century America. Among many others: 1) the argument between those foresters who seek the highest financial return from the forest, and those who, like von Salisch and myself, believe the aesthetic quality of the forest deserves a place, or a more prominent place, in the list of benefits; 2) the author's description (and opinion) of city tourists who visit the forest and complain about the visual disturbance caused by harvesting; 3) the scenario in PART I, Section A, Chapter 5, quoted from von Wildungen's *Hunter's Evening* (1815), describing an encounter between a pair of young idealists (a painter and a poet) and a forester, who are so far apart in thought that they really can not communicate to one another their respective opinions on the quality of the forest surrounding them. These problems and opinions presented by von Salisch are as real and as current as this year's calendar.

The translation presented several difficulties. One, deciphering the old-style alphabet was much more of a problem for me than for Doris, but even she had problems at times. A more serious problem was the antiquated terms as well as the old meanings of current words. There were numerous situations in which we were unaware of the change in meaning until, having read a translated passage, we realized it really didn't make much sense. So then we consulted Fluegel's *Woerterbuch,* 4th Edition, 1894; without it, our cause would surely have been lost.

Probably the most serious problem, and one which we can only hope to have resolved, is the forestry jargon, most of which doesn't appear in Fluegel's dictionary. Fortunately, Dr. Klaus Steinbeck, my colleague on the forestry faculty, loaned us his German-to-English dictionary of forestry terms published in 1939. Even so, a few terms were either missing or had no real meaning in modern American forestry. Some of these terms deserve to be mentioned before the reader begins the book, because our choice of terms may be misleading, particularly if the reader is at all familiar with 19th-century German forestry.

German foresters of the period divided their forest into *Jagen,* what we would call compartments; no problem here, although *Jagen* were quite small by American standards. The problem comes with their practice of bounding—thus separating—each compartment by *gestellen. Gestellen* were cleared strips between the compartments, and the closest comparable English word is firebreak. We used firebreak, although there was no indication that that had been their purpose; we could not definitely determine their purpose, although at least some were used for access to the timber within the compartments. On my recent visit to the forest that von Salisch knew as the Katholisch-Hammer, several, but not all, the compartments were bounded by cleared strips, and many of the strips were used for access.

Freie Anlagen was translated as park-like landscape, from its literal definition and the context. *Knick* is apparently an archaic or provincial word for hedgerow, which could be confused with *hecken*—hedges. *Femelschlag,* according to the forestry dictionary, is shelter-selection cutting, a practice either unknown in America or called by a different name; in either case, the distinction between it and *Plenterwald* (selection forest) is not clear. *Absaumungen,* or sometimes *Randabsaumungen,* was determined to be clearcutting in narrow, advancing strips—literally, cutting the border trees. Although we tried to stay close to the literal translation, we sometimes had to depend heavily on the context; thus, *Abnutzung,* or "using up," became "liquidation" when it was used in the context of cutting stands of timber. The author often used words with meanings that were unusual, judging from their position in the dictionary. Ferreting out what we eventually accepted as the correct meaning took a lot of digging and some trial and error in using the word in the proper context. We sincerely hope that we have captured the true meaning of von Salisch's work.

Another, less serious problem was what to do about the author's—and publisher's—lack of consistency in style and format. We eventually decided to standardize the chapter headings, spaces, captions, etc., rather than risk confusing the reader. In cases of compositional style, however, we tried to retain the author's work as much as possible, even to his unusual use of dashes with periods, and strange series of periods that had the appearance but not the function of very long ellipses. The actual language was very alive, almost conversational, and we wanted to keep that character.

The Index provided some unique problems, although we kept the editing to a minimum. Sometimes the word or the topic clearly did not appear in the text; in one case, a similar word had been used by mistake. In several instances, multiple synonyms were listed when only one was in the text. For example, both *Zirbelkiefer* and *Arve (Pinus cembra)* are in the Index, but only Arve appears in the text; likewise, Bienen (bees) and *Zeidelweide* (bee-keeping) both appear in the Index, but only the latter is in the text. We not only retained these duplicate entries, but we provided the reader with some additional entries, such as winter landscape and landscape, winter.

Names of trees, shrubs, and other plants are often a puzzle, even in the same language at the same time. Determining the true species of a foreign plant with a name that is 100 years old and possibly provincial as well is quite a challenge. Doris used Krussman's *Manual of Cultivated Broad-leaf Trees* and *Plant Communities of the Southeast* (unpublished lecture notes by Sam Jones and Darrel Morrison) as primary sources for names. These were checked and changed where necessary by reference to several other standard texts on dendrology and plant taxonomy. A few names were obtained from Fluegel's *Woerterbuch.*

Except for the Foreword, all of our additions to the text are in brackets, usually provided as synonyms to clarify the meaning. Items in parentheses were in parentheses in the original. German umlauts were replaced with an added e following the vowel, thus ä becomes ae, ö becomes oe, and ü becomes ue. Place names are in German or in English; in the Foreword, Polish or German (or English) names are used, depending on the period under discussion.

The decision to translate the Second Edition (1902) rather than the First (1885) or the Third (1911) was made rather by circumstances. The Second Edition was readily available by interlibrary loan from a number of libraries; it had been greatly expanded from the First, and had the advantage of the author's seventeen years of additional experience, as well as his opportunity to improve on the earlier effort. The Third Edition had not been changed as thoroughly, and it was not as readily available. The author forced us, however, to obtain the First Edition for its endnotes and references, since he did not repeat them in the Second Edition. These were included in the Appendix. A review of the Third Edition from a German professional journal has also been included; although the reviewer acknowledged improvement in the revision, there doesn't appear to have been any major changes.

Much of the Foreword has been condensed from Doris's comprehensive critique of the author's attitude toward nature and natural beauty, in light of his contemporary geographical, political, social, and philosophical situation. The relevance of the author's ideas in today's milieu, Central European as well as American, is discussed. I compare the forestry problems and issues of 1893 Silesia with those of 1993 America, noting particularly the uncanny similarity of many of the two periods. Also, with the generous help of the Polish Board for the Preservation of Historic Gardens and Palaces, and the foresters of the Milicz (Militch) Forest District, I describe the present condition of the author's former residence and estate, and the nearby public forests which he frequently used as examples in the text.

Walter L. Cook Jr.
Doris Wehlau
March, 1994

Foreword

The More Things Change . . .

In this foreword, I will comment on two aspects of change that have occurred in the years since Heinrich von Salisch wrote *Forest Aesthetics.* First, the changes in the political, social, and physical environment of Postel (Postolin).* This information has been accumulated from several sources. A major source is Doris Wehlau's *Historical Background and Today's Relevance of the Book "Forest Aesthetics" by Heinrich von Salisch, 1993*, a 131-page unpublished paper submitted in partial fulfillment of the requirements for the Certificate of Environmental Ethics, University of Georgia. Much information came from papers and maps provided by the Polish Board for the Preservation of Historic Gardens and Palaces, and translated from Polish to German by Irina Popiashvili from the Republic of Georgia. Additional information was obtained during my visit to Poland in August and September, 1993. While there, I was treated by Tomasz Zweich, Deputy Director of the Board, to a visit to the estate (now a park) of Prince Hermann von Pueckler at Bad Muskau, which now straddles the present border between Germany and Poland (the Polish side is called Leknica). Tomasz Zweich and Jakub Zemla then acted as interpreters during a tour of the estates at Postolin and Postolin-Karmin with Ewald Ranoscek, forester in charge of the former von Salisch properties. On the following day, we were guided by Marek

*The German or Anglicized place names will be used when referring to the period of German control, with Polish names in parentheses; in portions of the foreword that relate to present conditions, Polish names will be used.

Grobelny, forester and native of Milicz, who spoke excellent English. Linda Robertson, an American intern with the board and former classmate of Doris's at Georgia, also accompanied me on all the tours and helped immeasurably to make the trip a success.

The other aspect of change was the original instigation for the translation—to assess the similarities and differences in the relationship of forest aesthetics and timber (economic) management between the latter part of the 19th century in Silesia and the latter part of the 20th century in America. As the subtitle of this foreword implies, the similarities are greater than one might expect. Again, Doris Wehlau's paper is the major source of information.

The topography and soils of Postolin and vicinity were formed by the last continental glacier. What appear to be moraines are interspersed with relatively flat terrain. Numerous ponds, lakes, and swamps lie immediately to the north. The soil is loamy sand or sandy loam, with the moraines containing finer textured soil. The coarser soils support Scots pine *(Pinus sylvestris),* the most widely planted species. The finer soils with better moisture holding ability favor Norway spruce *(Picea abies),* beech *(Fagus sylvatica),* and other broadleaf species of trees and shrubs.

Heinrich von Salisch was born in 1846 at Postel, a tiny farming hamlet about 50 km north of Breslau (Wroclaw) in Silesia (Slask), then part of Prussia (Map 1). The nearest town of importance is Militsch (Milicz), about nine km to the northeast and Kreistadt (county seat). Heinrich's father, Rudolf, had bought the property in 1826; the palace was built in the early part of the 19th century, possibly after von Salisch's purchase. Rudolf and his family were members of the Prussian nobility, and maintained a feudal relationship with the peasantry. Heinrich studied forestry (he hints that his father was a forester or landscape architect), but apparently he became disenchanted with the state administration and did not take the exam for head forester. He had been experimenting with aesthetic forestry for ten years before publishing the first edition of *Forest Aesthetics* in 1885.

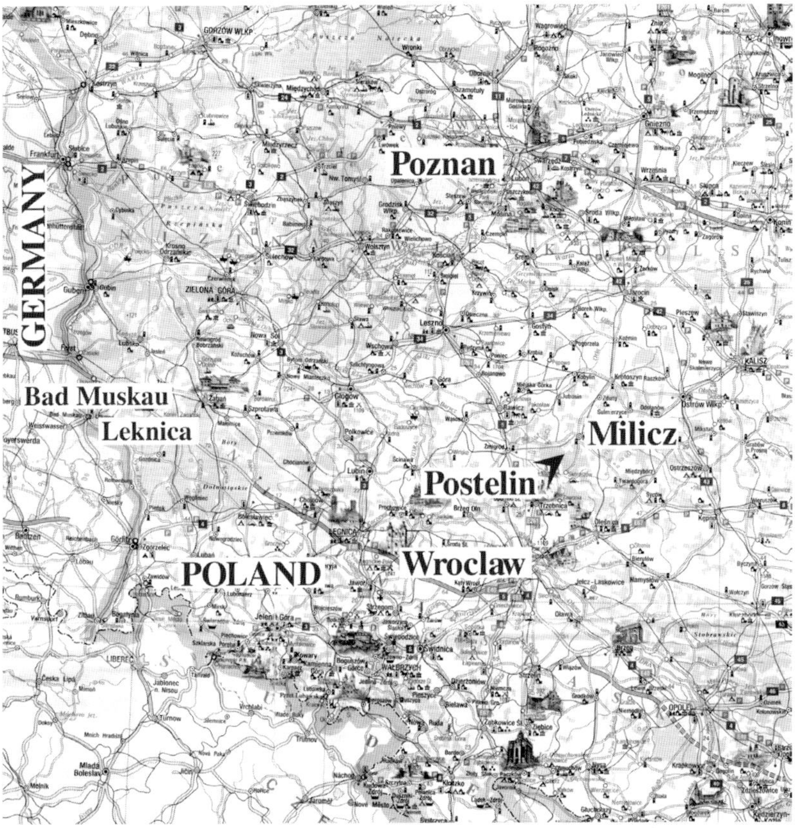

Map 1
Map of southwest Poland. Postelin, homeplace of author Heinrich von Salisch, is nine kilometers southwest of Milicz (Militsch).

Upon his father's death, Heinrich inherited the estate, including 665 hectares of forest. In 1888, he bought the neighboring estate at Postel-Karmine (Postolin-Karmin), about one km north of Postel. In 1893, he became a member of the Reichstag (the federal parliament), a post he held at the time of the second edition in 1902. He published the third edition in 1911 (see review in Appendix). He died in 1920 and is buried in a separate plot in the village cemetery; his wife, Susanna, died in 1926 and they share a common headstone. The last owner of the estate, in 1945, was Rudolf von Salisch, possibly a son, grandson, or nephew.

The latter part of the 19th century and early 20th century saw the last vestiges of feudal society in Silesia. Prussian nobility, many of whom had held the same property for many generations, were the landlords for the German and Polish farmers. The land was originally settled (c. 800) by Poles, but over several centuries, Germans moved east and intermingled with the indigenous Slav settlements, with each group retaining its own culture. With each period of agricultural depression, land was abandoned and taken up by the wealthier landowners, thus creating a land-rich class known as *Junkers*. The landless farmers became serfs, working for the nobility.

Meanwhile, the political power had gone from Polish to Polish-Lithuanian to Austrian to Prussian, with a brief period of French during the Napoleanic Wars. A castle in Milicz was built in the 14th century, ruined and rebuilt, and ruined again in 1795 (Photo 1), a testament to the repeated wars of the Central European powers. After that, Poland had ceased to exist as an independent nation, having been divided by Prussia and Russia, although three million Poles living under German administration had kept their culture, language, and Roman Catholic religion.

Rudolf von Salisch was apparently a *Junker*. The estate at Postel included several hundred hectares of forest. (The author gives two figures—"less than 1000 hectares" and "665 hectares"—but says nothing of the amount of cropland and pasture that was farmed by his tenants.) Heinrich added to the holdings by buying Postel-Karmine, containing an unknown area of land and forest. After World War I, Poland was restored as an independent nation, with the border being drawn only a few kilometers north of Postel and Militsch. Nothing is known concerning the estate following the war, except that the von Salisch family remained at Postel.

Photo 1
"The ruins of Gothic castle built by Prince Olesnickich in the second part of the XIVth Century. During Hus's Wars the castle was burnt. It was reconstructed in XVIth Century, when it gained its Renaissance style. It was partially burnt in 1797 and since then abandoned. The castle was surrounded by the moat and wall, behind which was located the courtyard."—Sign on the castle wall, translated by Jacek Siry, Polish graduate student in the Warnell School of Forest Resources, University of Georgia, Athens, Georgia.

World War II began with Germany invading Poland, ostensibly to re-unite ethnic Germans living in Poland—the same rationale used for occupying western Czechoslovakia a year earlier. As the tide of war turned in 1944–45, Russian troops entered Postel and leveled the von Salisch palace, ostensibly because one of its residents had been a Luftwaffe pilot. They also dynamited the local school building and burned the bakery, for no apparent reason.* The palace at Postolin-Karmin was not destroyed, although it has been abandoned and is in a very dilapidated condition (Photo 2). A four-inch birch is growing out of one wall, and its upper floor has partially dropped to the first; the roof is in poor condition, but the walls are still sound, and one can get an idea of its former beauty, and by extension, the appearance of the now ruined palace at Postolin.

Photo 2
Ruined palace (manor house) at Postelin-Karmin, about one km from Postelin, and once owned by the author.

With the end of the war came the western movement of Poland's eastern boundary with Ukraine and the western movement of the western boundary with Germany, to its present location along the Odra (Oder) River and its tributary, the Nysa (Niesse). All of Silesia became Polish Slask, and its German inhabitants, many of whose families had been there for centuries, were removed to the west of the border. Poles from the eastern area around Lwow (Lvov) and Briest (Brest), which had been ceded to Ukraine (then part of the Union

*This information came from a resident of Postelin, name unknown, during a casual conversation, through forester Marek Grobelny as interpreter, during my visit to Postolin in September, 1993.

of Soviet Socialist Republics), were moved to Slask. The area is now more ethnically pure than at any time since the 13th century.

Photo 3

Sign at entrance to von Salisch's park in Postelin (dot at major intersection on map), erected by Polish foresters from the Milicz office. Numbered trees are identified in Polish and Latin.

The von Salisch property was divided among many new Polish owners, but the forests were joined with the former Prussian state forests and are now national forests, managed by Polish foresters using the same district office in Milicz that had been used by the German foresters. The sites of the palaces at Postolin and Postolin-Karmin and their respective parks are also under the care of the Milicz foresters. Forester Ewald Ranoscek is in charge of the two sites, and he has mapped and inventoried the 5.2 hectare park at Postelin, and has erected a map board at the entrance which shows the species and location of each notable tree (Photo 3). Nature has reclaimed the lawns, the building site, and most of the roads, but many of the trees planted by Heinrich von Salisch and his father are in excellent condition (Photo 4). Although the routed wooden labels were mostly missing, Mr. Ranoscek identified the tallest tree in the park, a Douglas-fir *(Pseudotsuga menziesii)*, and the largest diameter, a European alder *(Alnus glutinosa)*, which was over four feet in diameter. The allee trees, which von Salisch described as being along his drive from the village to the palace (Page 253), could

Photo 4
"Pyramid oak" (*Quercus robur fastigiata*) in the park at Postelin.

not be identified. Of the remaining original trees, the English oaks *(Quercus robur)* and blood beeches *(Fagus sylvatica var. purpurea)* were the most impressive with their massive size and picturesque form. A hornbeam (*Carpinus betulus*) showed signs of having been pollarded many years ago (Photo 5). A surprise was bald cypress *(Taxodium distichum)*, which, at 51 degrees north latitude and in the cold middle of the European continent, is far from its home in the warm swamps of southern United States.

Photo 5
Hornbeam *(Carpinus betulus)* in the park at Postelin.

Given the number and diversity of species von Salisch and his father were able to establish, the natural reproduction is composed of surprisingly few. Prominent were several species of *Tilia* (nearly all the older reproduction, now pole-size, was *Tilia*), *Robinia pseudoacacia,* and *Acer pseudoplatanus,* both the species and the red-leafed variety. Others noted were *Prunus serotina, Acer campestre, Acer platanoides, Ulmus laevis,* and *Fagus sylvatica.* Much of the lower understory was shrubs, including *Sambucus, Sorbus,* and *Callicarpa.* The only conifer reproduction was of common yew *(Taxus bac-*

cata), although several tolerant conifers were seen among the original planted trees. The author von Salisch would have been happy to know that one of his favorite species, English ivy *(Hedera helix)*, had been quite successful, having covered large areas of ground in an impenetrable mat, nearly hiding a hedghog (Photo 6) and in attaining a diameter of about 10 centimeters, climbing on an English oak. Other sites with trees planted by von Salisch included the park at Postolin-Karmin, and the family cemetery plot, where several large ponderosa pines *(Pinus ponderosa)* were found.

Photo 6
Hedgehog *(Erinaceus europaeus)* trying to hide in the rampant English ivy *(Hedera helix)* in the park at Postelin.

The sight of the overgrown park, the scum-covered pond, and the pile of rubble that was once a fine palace are all somewhat depressing, although the diversity of species provides an arboretum-like quality (the students at the forest technician school at Milicz likely use it for that). This should not be surprising, given the ferocity of war and the relentlessness of nature in reclaiming an abandoned site. What was certainly a surprise, and a welcome one, was to see that Johanna's Height, the stone hunting tower built by Rudolf von Salisch in 1849–50 and named for his wife (Heinrich's mother), was still standing (Figure 42, Page 222). One would have guessed it would be used in directing artillery or in general observation, and as such, would have been a prime target for an enemy. But it survived all the wars intact, only to have its wooden interior gutted by fire in the late 1940s. The Polish foresters restored it beautifully, added five meters to its height, and used it as a fire lookout and hostel (Photo 7). At the junction of its approach road, they have erected a sign in the same style as the one at Postolin, giving the history of the tower. The beautiful 150-year-old beech forest along the approach road is the same stand as that which von Salisch thinned by his so-called Postel thinning method (Plate VIII, Page 190). Ironically, his objective was to prevent the stand from becoming "transparent" by

Photo 7
Johanna's Height (hunting tower) built by the author's father in 1850 and restored by the Milicz foresters.

encouraging understory beeches; today, the forest floor is covered with forbs, under a high canopy of mature beech of impressive stature, but no understory is present (Photo 8).

Photo 8
By decree of the Polish Ministry of Forestry, this 24-hectare stand of 150-year-old European Beech *(Fagus sylvatica)* surrounding Johanna's Height has been placed in a reservation. It is likely that this stand was where von Salisch practiced the "Postel thinning method."

The tour of the former Katholisch-Hammer Forest, led by forester Marek Grobelny, revealed very little physical change. We followed the pre-war German topographic map that showed each compartment *(Jagen)* and firebreak *(Gestell),* as well as the roads running through the compartments (Map 2). The compartments, firebreaks, and roads were all there, as if time had stood still. Even the squared granite markers at the compartment corners were in place, invariably on the northeast side of the intersection; the Polish foresters had merely changed the compartment numbers (Photo 9). We first looked at Compartments 139 and 140, which the author referred to as Paradise (Page 147 and 154) and Compartment 149, called

Pickel's Warm Beer (Page 148). These names were also listed on the topographic map, along with their numbers. We could see nothing outstanding about these sites that would explain their names.

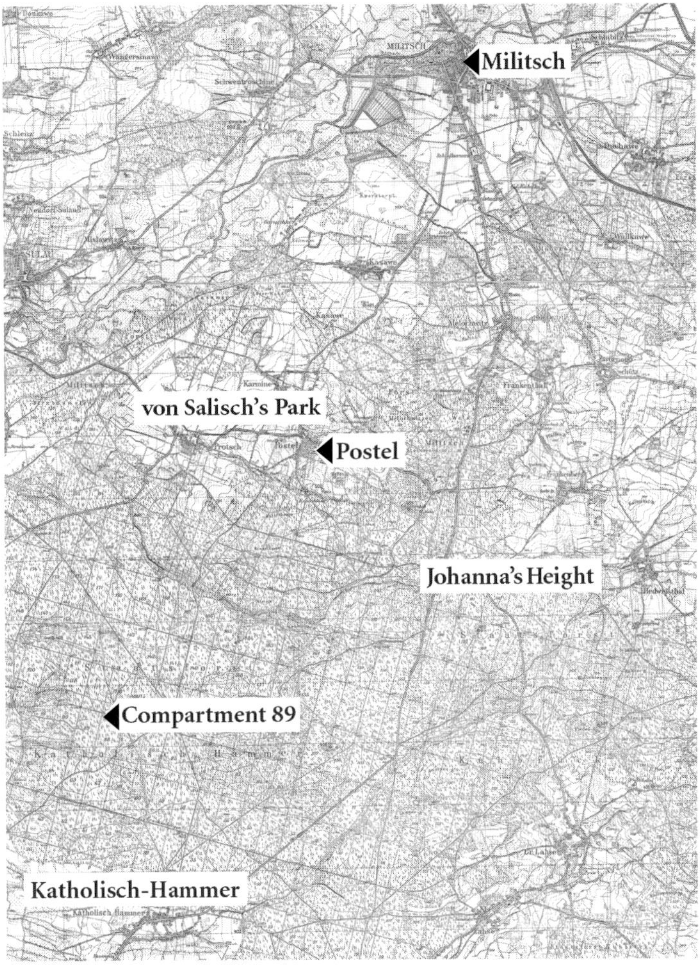

Map 2
German map (c. 1928) of Postel and vicinity, showing site of author's palace and park, village, his surrounding forest, and the neighboring state forests, including Katholisch-Hammer District. Arrow points to intersection of Compartments 88, 89, 116, 117. See Figures 28 and 29, Pages 144 and 145, and Photos 9 and 10.

Photo 9
Granite marker at intersection of four compartment boundaries in the Lasowice Forest, formerly the Katholisch-Hammer State Forest District.

Photo 10
Compartment 89 (German number) being cut at the age of 106 years. The Scots Pine *(Pinus sylvatica)* are likely the same trees planted after the clearcut shown in Figure 28; the two photos were taken from the same location.

And then we went to the intersection of Compartments 88, 89, 116, and 117. Figure 28, on Page 144, is a photograph taken from this intersection, showing Compartment 89 in the process of being cut, its 100-year-old trees having reached the established rotation age. Now, in 1993, the compartment was being cut again (Photo 10). A ring count on the stumps revealed an age of approximately 106 years. Those trees *(Pinus sylvestris)* were very likely the same ones that were planted following the clearcut shown in the photo on Page 144!

We ended our tour by visiting Klein Ujeschuetz, a nearby hamlet on the Trachenberg Line, a major firebreak that von Salisch referred to on Page 213. We looked for any trace of the "cypress-shaped junipers" that bordered the forester's private plot on the Trachenberg Line. No junipers were found, and the site of the forester's house was a pile of rubble.

Changes in Aesthetic Philosophy and Its Application to Forestry

Due to the intermingling of philosophical, political, and forestry statements in the book, as well as to occasional inconsistencies and contradictions, it is not easy to give a short overview and analysis of the author's arguments. Because the application of aesthetic principles to forestry is based on his philosophy of beauty as well as the influence and constraints of economic forestry, we will begin by briefly summarizing his thoughts on aesthetic philosophy. Then we will compare his applications of that philosophy to the practice of forestry in the Silesia of a century ago with the current situation and practices in America.

Von Salisch states that "the human does not enjoy beauty with his senses, but with his intellect" (Page 11). He indicates that humans have an ideal abstract interest in the existence of beauty that is neither dependent on the actual experience of a work of art, nor connected to a certain interest. The aesthetic pleasure is pure, without any interests. Salisch takes the position that beauty is a quality intrinsic to things that is totally independent from the existence of humans as viewers. The value of beauty does not depend on its rarity, nor on other economic laws of the free market. This concept is all the more remarkable when one considers von Salisch's profession, his social status, and the booming capitalism of the times. He uses the example of a potato and a rose. Although the potato is very useful, not many people would think it is beautiful; on the other hand, the rose, which has no use whatever, is universally considered to be beautiful.

Although von Salisch does not refer to George Santayana, he likely would have agreed with this philosopher who was his contemporary, concerning the relation between beauty and utility. According to Santayana, a thing need not be useful to be beautiful, but if it is supposed to be useful but isn't, perhaps because of some flaw of design, then it cannot be beautiful (Santayana 1955 [1896]). This could explain one of the more troublesome ambiguities of the book—while von Salisch believed that beauty was independent of human senses and economic value systems, he was adamant that management activities that promote the aesthetic quality of forests must not sacrifice all of the income from timber production. Forests that had no economic value—wilderness areas, for example—were not to be tolerated. But throughout the book, one of his most repeated arguments was for the inclusion of aesthetic value in the equation *alongside* the market value of the material products of the forest in determining the rotation and the net present value. He believed that aesthetic benefits could and should replace some, but not all, economic benefits.

Von Salisch held the opinion that the "nature of beauty is based on harmony or unity in variety" (Page 25). The best relation between parts and the whole is given, if the unity is kept while at the same time there are strong divisions. This would be the case when the parts stay incomplete and they need each other for completion. This relation of the parts to the whole is one of the most important principles in design. It is also one of the basic concepts of ecology, where the whole is supposed to be more than the sum of the individual parts, an idea that did not exist in von Salisch's time. He describes in detail the relations of the golden section as being the most harmonious and at the same time "natural" proportion of parts towards the whole and each other. Further properties of beautiful things, according to von Salisch, are the apt arrangement of the harmonious sections determined according to the rules of the golden section (a.k.a. golden mean).

Decisions and statements on beauty are always evaluative. If beauty is a positive value statement, it may be connected to other positive statements that are true at the same time. Human values, however, are a cultural factor (except basic human needs) and they may differ significantly in different cultures. Thus, our (American) values in 1993 differ in some respect from the values of the Wilhelmian society of Germany in 1893. That cultural values are important, and that they change over time, is illustrated by von Salisch's list of examples of beautiful objects (Page 19): "Cologne Cathedral, a butterfly, the Ninth Symphony of Beethoven, a forget-me-not, the marchpast of a cuirassier squadron, Schiller's *Wallenstein,* a sunrise, an oak, Shakespeare's *Romeo and Juliet,* the Prince Bismarck, [and] an orange . . ." He expresses no doubt that "anyone" would agree that these items are beautiful; obviously, their beauty depends on the time, the culture, as well as on personal taste, which to a large degree is formed by one's culture.

One of the differences in values is in the emphasis on order. The author believed that ". . . everybody can contribute in a certain way to the enhancement of beauty, even if it were just by cleanliness and love of order" (Page 4). Here it can clearly be seen that the contemporary cultural values of a society do influence the judgment of beauty. Aesthetic considerations are not free from cultural influences, nor are they based on some "eternal intrinsic" qualities. The judgment of beauty is influenced by norms of the society, brought into the individual by education and social conditioning. "Cleanliness" and "love of order" sound like typically German values; they were emphasized in Prussian society in order to create discipline and increase efficiency.

The respect for order, neatness and cleanliness are still today important values in the lives of many German citizens, particularly older ones. And so it is in America, where the traditional values of hard work and responsibility are associated with cleanliness, neatness and order, especially in rural and small towns where forests and farms make up a major part of the landscape as well as the local economy. In a photo-preference study involving Upstate New York woodland owners, Hamilton (1967) found that scenes with thinned and pruned pine and spruce plantations, with ricks of cordwood from the thinnings, were much preferred. Unthinned and unpruned plantations or those with invading hardwood understory were disliked, but natural hardwood stands with a hardwood understory were liked. This is evidence that the owners appreciated the thinned stands as a sign of good husbandry, which is associated with beauty. The uncared-for (read unkempt, untidy, messy) stands lacked the order that they, being plantations, should have had. The invading hardwoods were seen as weeds which indicated disorder, whereas the hardwood understory in the natural hardwood stand was orderly—it was supposed to be there. In the southeastern United States, where the major forest crop is planted loblolly pine, foresters and many landowners strive for the "perfect forest"—a pine plantation with all vigorous, well-formed trees, and a clean understory; the more a stand approaches this ideal, the more beautiful it becomes. This is in total agreement with von Salisch's quote from forester Koenig's *Forest Care* (Page 13): "A forest in its highest forestry perfection is also in its most beautiful state."

This respect for order is surely related to utility; von Salisch equates "The Beautiful" with "The Good" and "The Useful." He states ". . . the consideration of aesthetic points of view protects against economic mistakes, because along with the striving for beauty, which leads to perfection, one achieves the good and therefore the useful at the same time" (Page 7). This seems to be spurious logic; the author apparently was trying, as he did throughout the book, to justify the expenditure of effort (and likely the sacrifice of part of the profit) for the sake of aesthetic values. Not only is his argument illogical, it also contradicts what he had said earlier, that aesthetic pleasure in the beautiful is free from any interest.

Whether one should consider it equivocating or merely an honest misunderstanding of terms, von Salisch provides still another argument to rationalize his defense of economic forestry while advocating aesthetic forestry. He states that aesthetics, "according to a more strict limitation of the term, has nothing to do with good and evil, with useful or useless, but is limited to what our senses transmit to us" (Page 24). This is actually the meaning of aesthetic pleasure in beauty in its real sense, meaning that the perceiving person is free from interest. Thus it has been defined and used by aesthetic philosophers. Salisch defends his own approach to the subject of beauty, which, as seen above, is closely connected with his considerations of goodness and usefulness for humans, with the pragmatic argument that otherwise he would not be able to find aesthetic principles applicable in the forest. He bases this on the statement that in reality, nobody is free from other influences in his aesthetic judgment, thus the real conditions are that subjective knowledge and experience cannot be ignored, and he wants to base his forest aesthetic principles on real life conditions. Because real life and the cultural preference of the contemporary society are like that, he takes them over as preliminary conditions for his aesthetic principles. The question then is, why has he discussed aesthetic theories at all, if they are just theory and not reality?

Von Salisch's theory of "idea-association" has been the subject of much research in the past twenty years. As he explains idea-associations, people will prefer or not prefer an object or scene, depending on what the person associates the object with in his or her life. Thus, if a person sees a clearcut, and he has read or been told that clearcuts are bad for wildlife, or that they will cause springs to dry up, then the person will likely not like the sight of the clearcut. On the other hand, if the person thinks that clearcutting will control the tick population, he will be more apt to approve of the practice. Anderson (1981) and others have set up series of photos of forest scenes, all from the same forest and all relatively similar. The photos were then labeled with suggestive titles, such as "commercial logging," "national forest," "national park," "wilderness area," etc. Although the photos were essentially the same, the viewers ranked them according to their perceived quality—the ones labeled "wilderness area" were liked best, etc. While von Salisch does not have much to say about the practical application of this phenomenon, it is a major problem facing foresters in present-day America. A couple of other examples will further illustrate the problem.

Shafer and Rutherford (1969) showed photos of uncut (essentially virgin) northern hardwood forests, and photos of similar forest which had been cut conservatively by the single-tree selection method. The viewers were not told about the cutting, which was not obvious to the layman's eye in the photos. A significant majority preferred the scenes of the selectively cut forest. Kaplan and Kaplan (1989) reported a test involving beetle-killed pines along the Blue Ridge Parkway, still with their distinctive orange-brown foliage. Those respondents that were not told the nature of the pines preferred scenes containing them, but respondents who were told why the pines were brown did not prefer those scenes. Thus von Salisch's opinion (he agreed with Humboldt) that detailed knowledge of the object being viewed may increase our appreciation of it is very likely based in fact, although the knowledge can work both ways.

Von Salisch begins Section B—The Beauty of Nature—with this statement: "Natural beauty is particularly important for us foresters who live more outdoors than many other professions and have to guard and to care for a rich treasure of natural beauty" (Page 34). An axiom in the study of aesthetics is that natural beauty, if in fact it exists, cannot be as beautiful and as perfect as a work composed by an artist. The author quotes Hermann, "The individual thing in nature is, on the average, always less perfect in regard to its appearance than a work of art" (Page 35), and then vehemently disagrees with him. He says in rebuttal, "This opinion is rooted in the perception, quite correct in itself, that the artist can never simply imitate nature. While the artist redesigns for the specialties of his art the material nature brings to him, it comes with the vanity which exists even in great men that he thinks of his own work as more perfect than the original" (Page 36). In taking this position, von Salisch agrees with every forester I have ever known. Foresters are surrounded by beauty in the everyday course of their work, and to suggest that they do not recognize beauty does them all a great disservice.

This belief by foresters that nature is beautiful is likely related to the loving attitude foresters have for the forest. According to von Salisch, who quotes Pfeil to make his point, there is a certain kind of love which should be instilled into the young forester. This love should not be based on economic profits, but in ". . . the intimate interest in the prosperity of the trees and forest, the striving to enhance it for the sake of the trees themselves. . . ." (Page 8). One hundred years later, young foresters are still being instilled with a love of the forest and a desire to enhance it for its own sake. Is it possible that some foresters have lost this idealistic notion, and their single-minded attention to the economic return has cost them the respect, admiration, and trust of the lay public?

Certainly, there is a difference between the forestry of 1893 and 1993, but most of the difference is due to the combination of greater knowledge of the forest and the advances in technology—in 1893, there were no computers, no chainsaws, and no tractors. With our modern knowledge and equipment, we can do a much better job of managing forests; on the other hand, when we make a wrong decision or misapply the knowledge, the result is a mistake of much greater magnitude. Von Salisch wrote of the healing hand of nature that could balance many mistakes, but the mistakes that could be made then were of a magnitude that *could* be healed. Our mistakes will be a long time healing.

Another difference that should be noted is related to the

knowledge explosion. In 1893, von Salisch was among the last of the Romanticists. He read and was inspired by philosophers, aestheticians, poets, painters and other artists. Some of the people he most frequently quoted or referred to were Gilpin (landscape architect, painter, and writer), Krause (philosopher), Pueckler (nobleman, landscape architect, idealist), Goethe (poet). The leaders among the foresters who are currently rebelling against the "establishment" are mostly ecologists and other scientists who are concerned that serious and irreparable damage is being done to the forests in the name of efficiency, production, and profit. Most of the "rebels" are not Romanticists, or "tree huggers," as they are sometimes called. There are, however, a significant number of foresters and other natural resource managers and technicians who seem to have wedded Romanticism with ecology in a belief that humans are part of nature, and that we should behave as responsible members of the land community. Known as Deep Ecology, this movement springs from the writings of Thoreau, Muir, and especially Aldo Leopold.

Perhaps related to the current concept of "stewardship" is von Salisch's belief that God has given the earth to humans for use and cultivation, but with it also the duty to do good, not to do evil, and to increase and shape the beautiful. It is a part of the art of cultivation of the earth to shape it into a beautiful home for mankind. Cultivation of the earth, that is to make it useful for human purposes, thus is a totally good thing that does not even need to be discussed.

In one sense, von Salisch was a rebel forester, but in another he was a pioneer. Certainly, writing a book with the title, *Forest Aesthetics* was nearly unprecedented. He mentions the book, *Aesthetics in the Forest,* by von der Borch in 1824, but it could not find a publisher. Other major references are largely concerned with park landscapes rather than aesthetic treatment of commercial forests. Given the economic influence in German forestry during his time, the fact that he found a publisher for three editions, had reasonably polite reviews in the forestry publications, and had opportunities to present his ideas at various conferences, all point to a rather important and respected contribution to the profession. This prominence for the field of forest aesthetics would not be repeated in America until the peak of the environmental movement in the 1970s. John Curry, near the end of a 40-year career in forestry in New York in 1961, revealed his disappointment that no one had cared enough for the beauty of the forest, that forest aesthetics was almost unknown in America, that there was no course on forest aesthetics taught in any American forestry school, and that he knew of only one old book on the subject, written by a German forester many years ago (Curry 1961). Curry was obviously referring to von Salisch's book. (It was this article by Curry that sparked my career interests, and started me on the way to graduate school in 1967, where the faculty were also unfamiliar with forest aesthetics.)

Von Salisch regrets that the eagerness to do research and to write is much less applied to forest aesthetics than to other fields of forestry, e.g., the calculation of net profits and interest. One hundred years later and several thousand miles west, the same statement can be made. Foundations, government agencies, and other usual sources of research funds are rarely available to the scholar who would do research in forest aesthetics. As an example, the U. S. Forest Service turned down a request for a small fund to aid in this translation project; likewise, the trip to investigate the author's estate site at Postolin was only partially funded, on the grounds that it wasn't "research." For a brief period in the 1970s, articles on forest aesthetics would appear in the (American) *Journal of Forestry*. To date, no textbook has been published, and I am not aware of a forestry school other than at the University of Georgia that offers a course in aesthetic forestry; even that one is very poorly attended.

The basic issue of the book is whether state (public) forests should be managed solely to produce the greatest possible economic benefits, or should aesthetic values be one of the objectives. One major conflict between these two objectives lies in the rotation age. Economic value always peaks before aesthetic value, partly due to the interest rate and partly due to the fact that trees slow their growth rate appreciably before they reach their full aesthetic potential. Thus, income would be sacrificed by leaving the tree grow to this somewhat overmature state; in addition, the chance that something dire might happen to the tree increases with increasing age. Von Salisch counters the economic argument with at least two plausible arguments.

First, von Salisch quotes a fellow forester, Guse (Page 5-6), to argue that the state spends huge sums of money adding decoration to public buildings such as railroad stations, post offices, and administrative buildings—all in situations where those in power can impress the people. Why should the people's forests be an exception? People like to admire beautiful forests as much as they do beautiful buildings. Guse argues that the poor worker gets his spirits lifted by walking through a beautiful forest, after a day's work. The worker cannot afford costly decorations in his home, but the forest provides beauty for him for free. This is nearly identical to the rationale used by Frederick Law Olmsted in explaining his design of New York's Central Park several years earlier, as well as by most other landscape architects of the period. A third argument was that if people like the forest (and they would like a beautiful forest before they would a forest that was not beautiful), they would be more protective of it, defending it from those who would destroy it. This is related to the argument that pride in a beautiful forest would also increase pride in one's country; patriotism was a major part of German culture in the late 19th century, and continued until the end of the Second World War.

These arguments in favor of a beautiful forest are not heard in late 20th century America. Now, natural beauty is equated with naturalness—lack of obvious signs of human interference, manipulation, or intrusion. It seems we are condemning ourselves for having ruined and ravished our continent, and look to the opposite extreme—"untouched wilderness"—as the ideal, where "nature" reigns unimpeded by human "management." One aspect of von Salisch's philosophy that is also a tenet of the Deep Ecology movement is humility. We should be humble towards nature if we are to create beauty in the forest, according to the author.

One of von Salisch's arguments is used in defending current efforts to limit forest harvesting: forest beauty is an economic factor in areas where tourism is important (Page 222-224). No motel owner wants the forest across the road to be clearcut, or in any way have its beauty (naturalness) diminished. Even the U. S. Forest Service recognizes that tourists don't want to see destruction up close, and in some cases at least, modifies its harvesting and silvicultural practices in sensitive areas. However, von Salisch's argument that *all* the forest should be given the same considerate and conservative management, on the grounds that we don't know where the next new public road will be cut, would be as unpopular today as it was in 1893. On the other hand, his argument that as a professional forester, he knows more than the lay public and therefore his opinion is "more valid than that of many tourists," is quite popular among my contemporaries; it is not used to justify aesthetic practices, however, but to justify questionable silviculture in the search for greater profits.

In von Salisch's time, the use of the forest was mainly for timber and hunting. Recreational use was known, but not in the amount or style of today. Without the mobility provided by the automobile, the increased leisure time, and the affluence with which to enjoy it, most recreation was local, and limited to contemplation and appreciation of nature. Romantic idealism, as interpreted and modified by the landscape artists and architects, was adopted by the common people. The virtues of "nature" were extolled, but it wasn't the nature that many of us associate with wilderness today. Von Salisch mentioned wilderness, or rather ancient forests, but he didn't approve of them. He didn't think the people would enjoy climbing over dead and rotten tree trunks; besides, he felt wild, unmanaged forests constituted a waste. He firmly believed that forests should be beautiful *and* useful—no exceptions.

Today, the situation has profoundly changed, both in Europe

and America. The national park idea, which had just begun in America in 1872, and to which von Salisch referred with a hint of envy, has since spread around the world. State and provincial parks have added to the outdoor recreation resource base. The greatest change, perhaps, is in the use of state and federal forests for timber production *and* recreation; this is true in both the United States and Germany. In Germany, near urban areas, the silviculture is chosen deliberately to enhance the recreational (aesthetic) satisfaction—in other words, a high forest is maintained by mostly shelter-selection cutting, and monocultures are regarded as less desirable. In the more remote areas, clearcutting and monocultures are still seen, but on a smaller scale than in America, and the harvesting is done with smaller machines. Thus today, von Salisch's ideas of integrated forest management have been put into practice, but on a scope he could not have imagined. The U. S. Forest Service, driven by public outcry against the excesses of economic forestry and concern about ecological destruction, has recently made some major policy changes that may eventually emulate the German formula.

An important difference between the two countries is that, with vastly more land, the U. S. has placed large areas into wilderness areas, a status designed to prevent *any* manipulation of the forest, where only trails are provided for recreating visitors. Another difference is the proliferation of mostly small, protected natural areas, owned by the state or local government, or by private conservation organizations. A 1988 survey revealed a total of more than 5,000,000 acres (2,000,000 hectares) in nearly 4000 separate parcels (Cook and English 1988). Many of these areas provide the contemplative, nature-oriented type of recreation von Salisch knew and enjoyed.

At the time von Salisch was writing his book, the science of ecology did not exist. It is interesting to note, however, how many of his recommendations that were based on his aesthetic sense are currently advocated, but for environmental or ecological reasons. For example, he recommended maintenance of the natural groundcover and understory of a forest, or to allow for an understory and middlestory to develop while the forest matures. (He did not like "transparent" forests!) He preferred to allow small groups of natural reproduction to remain, rather than eradicate the group because it did not fit the pre-conceived idea of pure and orderly plantations. (This "purification" was practiced by the U.S. Forest Service, in the name of efficiency.) He was adamantly opposed to the planting of exotic species in production forests, whether for beauty or economics; he did approve of exotics in parks, and he had a veritable arboretum surrounding his palace. His prejudice against foreign species apparently was based on a form of patriotism; he felt there were plenty of German species available, and that foreign species just didn't fit into the German landscape. In America, there is a very recent movement opposed to foreign species, but this applies more to landscape plantings and natural area preserves than to production forests. Certainly, with the vastly greater number of native species in America, there is less incentive to plant exotics.

Forest roads were another item about which von Salisch had strong opinions. Again using aesthetic reasons, likely adapted from Gilpin ("all straight roads displease"), he argued against straight roads and compartment boundaries (which were marked by firebreaks or roads). The German topographic map of the Postel-Katholisch Hammer area (Map 2) shows very dramatically that compartment boundaries and roads on his estate forest were quite irregular, while those of the state forest to the south (Katholisch Hammer District) formed perfect rectangles. He advocated allowing the natural landforms to determine road layout, something that is promoted widely by foresters in this country; the difference is that the forest roads in America that are not designed by foresters are nearly always located by chance, or at the whim of the logger, with expediency being the only criterion.

Although ecology did not exist as a science in 1893, forestry had been practiced and researched in Germany and other parts of Europe for at least one hundred years, and some say for several hundred years. It is known that von Salisch had been influenced by this accumulated knowledge, since he had studied forestry at the university. It is also known that he had been influenced by the Romanticists and the philosophers of the period, since he frequently quoted them in the book. But there is another source for his advocacy of aesthetic practices. His reasoning was based on a deep understanding of natural processes and a respect for natural forms and functions. He had an inner sense of nature which comes from having a close association with the forest and the land. Many foresters and farmers have this sense; Aldo Leopold had it to a great degree, and he wrote it all down, just as von Salisch did before him. The difference is that Leopold dared to go farther in denouncing the economic bias in natural resource management, and to formulate a new land ethic, based on his inner sense and bolstered by his ecological knowledge. Leopold, in 1948, was just as far out of step with the mainstream philosophy as von Salisch had been in 1902; but note that Leopold remained nearly unknown to the public until the Second Conservation Movement (1962–present) discovered and adopted his writings as its guiding principle. It would be interesting and possibly enlightening if we could determine whether Leopold had read von Salisch's book.

The discussion of conflicts between economic benefits or necessities and environmental and aesthetic benefits is certainly not unknown to us today. We are still living in a worldwide political order that is based on economics as the main criterion. Most decisions concerning land and natural resource use are made by the standard of monetary or materialistic gains or losses. Thus, the conflict between economic reasoning—the maximum profit—and reasoning in terms of nature's beauty and well-being is still the same as it was in von Salisch's time. The arguments and the vocabulary have changed, but basic conflict still lies in the fundamental personal attitude towards life and the world. But the living standards and levels of wealth are so much greater today that the whole process of maximizing profits is taking place on a higher level of magnitude. Greed and demands have grown; we want more and more and higher and higher profits. With the decline of the Christian elements of humility and modesty in people's world view, there is no limit left for human selfish striving for advantage. We need a new set of criteria with which to guide our relations with the forest and nature in general. Perhaps someone reading this book will be inspired to develop those criteria.

Walter L. Cook Jr.

References for the Foreword

Anderson, Linda M. 1981. Land Use Designations Affect Perception of Scenic Beauty. *Forest Science* 27(2): 392–400.

Cook, Walter L., Jr., and Donald B. K. English. 1988. Non-federal Wilderness, Wild, and Natural Areas in the United States—A Survey. *In*: Outdoor Recreation Benchmark 1988: Proceedings of the National Outdoor Recreation Forum. Tampa, Florida. January 13–14, 1988. USDA Forest Service General Technical Report SE-52. Pages 319–335.

Curry, John. 1961. Forestry and Forest Aesthetics. *Northeastern Logger* 10(5): 26, 36–37.

Hamilton, Lawrence S. 1967. Private Woodlands, the Suburban Forests and Aesthetic Timber Harvesting. *Forestry Chronicle* 42(3): 162–166.

Kaplan, Rachel, and Stephen Kaplan. 1989. *The Experience of Nature—A Psychological Perspective.* Cambridge: Cambridge University Press.

Santayana, George. 1955 [1896]. *The Sense of Beauty.* New York: The Modern Library.

Shafer, E. L., Jr., and W. Rutherford, Jr. 1969. Selection Cuts Increased Natural Beauty in Two Adirondack Forest Stands. *Journal of Forestry* 67(6): 415–419.

PART ONE

Fundamental Principles of Forest Aesthetics

SECTION A. Preliminary Terms

Chapter One.

Term and Task of Forest Aesthetics. History and Literature of Forest Aesthetics as a Special Branch of Forestry Sciences. Necessity to Study it.

1. What to Understand by Forest Aesthetics

Forest aesthetics is the teaching of the timber forest's beauty. It is supposed to show where its beauty lies and how to take care of its beauty.

I will call forestry art the forestry economy that is done under aesthetic points of view. Forestry art is a branch of **landscape art,** whose purpose it is to shape the earth into a beautiful home place of mankind.

Landscape art itself is a branch of the **general cultural arts** on the earth.

In setting up these definitions, I follow the aesthetics philosopher Karl Christian Friedrich Krause, whose last work, called "*The Science of Landscape Art,*" was written in 1832, but published only in 1883.

According to Krause, several arts are part of landscape art: the art of building itself (architecture), the art of gardening, the art of forest cultivation, the art of crop cultivation and the art of meadow cultivation. Krause, however, thinks of healthy, robust, beautiful and happy people as being the most beautiful adornment of a country.

Concerning the art of forest cultivation, Krause says: "Regarding the art of forest cultivation, its main topic is the planting and growing of forest trees and forest shrubs, **for use as well as for beauty and pleasure,** and the nurture of forest wildlife. It also includes the construction of structures **which are necessary for use and beauty of the forest,** like houses, trails and roads, meadows and gardens and the necessary water reservoirs and water supply utilities."

The emphasized lines above are remarkable due to the sharp limitations they contain. Forests which are managed solely for beauty and pleasure, but not for timber use, are not part of the field of forestry art (or art of forest cultivation as Krause calls it); however, houses, roads, gardens and so on, which are **necessary** for the use and the beauty of the forest, are part of the field of forestry art.

We now have set forestry art as a branch of landscape art beside the art of gardening, but we need to warn of a probable error. For there is an aesthetic section which defines **idealizing nature** as a task of the art of gardening. Besides several other authors, Dr. K. E. Schneider defends this opinion, and he flatters himself of having initiated, according to his view, a "not yet existing science" in his "*Attempt for the Aesthetic Foundation of the Fine Art of Gardening,*" and of having filled a still existing gap in the system of aesthetics. Now, someone who agrees with Dr. Schneider and who believes that it is possible to idealize nature, might consider idealizing forest nature as a task of forestry art.

But how could man in all his weakness be capable of this task!

It is regrettable that Schneider's error does not stand alone. Already Goethe had to fight similar errings; he did it in his own way, in a short but appropriate poem. This poem is called **"Der**

Chinese in Rom" [**The Chinese Man in Rome**], and I would like to quote it here, because it is regrettably not well-known:

I saw a Chinese man in Rome. To him all the buildings
Of former and present times seemed to be tiresome and heavy.
Oh, he sighs, poor people! I hope they will understand,
How in some future time fine wooden columns will carry the roof's tent.
And that the educated eye's sensitive sense will be delighted
Only by slats, felts and golden ornaments.
Look, at that moment, I thought to see mentally one of these dreamers,
Who compares his aerial web with solid nature's eternal carpet,
Who calls the true pure healthy one sick
In order that he himself, the sick one, will be called healthy.

Some aesthetic philosophers have truly recognized the purpose of the art of gardening. For example, Vischer sets it as a goal of the fine art of gardening, "to idealize the **walk** on the picturesque colorful area."

If, however, it is not a task of the landscape gardener to idealize nature, then this certainly cannot be claimed to be a goal for foresters either; but nevertheless we do not lack an ideal. That is what I mean by the words: **The purpose of the art of forestry is to idealize economic forest management.**

Like architecture is related to the craft of bricklaying, so the art of forestry is supposed to elevate itself above the craftsmanship of manual labor in forest management.

With regard to forest nature, our task is a double one.

First of all there are **damages to be brought to an end, which man has caused.** Again I quote Krause:

"Art is the skill formed out of a force to create or to shape something significant, to put it into life and to truly maintain it in life. Man is able and capable to form bad things, evil things and things adverse to nature; but he ought not to do that. However, if he does so, it is not art, but it is sin and crime to create and to shape things adverse to nature."

How many "things adverse to nature" are there to be redesigned!

Furthermore, it is our task, as Krause says, **"to shape nature beautifully in new respects and relations."**

In explaining this, I point out that this requirement is fulfilled if the timber stands are grown with regard and relation to their future optimum utilization and in a beautiful way at the same time.

Thinned out, fungus-infested mature woods may be beautiful in a park, but they cannot adjust to the forest economic "respects and relations." These, therefore, have to be substituted by younger trees, which promise utilizable timber.

The difference between forest and park woodland will be discussed in detail in the applied part, therefore I do not wish to anticipate it here.

2. It is Necessary to Consider Aesthetic Points of View Everywhere in Forest Management

"The beautiful is absolutely meant to be, therefore there is as well an absolutely obligatory demand that rational creatures shape the beautiful."

This remark of Krause can be proved by religious reasons and in the philosophical way. The reader might want to read this in the quoted source. Here the remark may suffice that this quoted sentence deserves particular consideration in the large field of landscape art due to the points of view of Christian brotherly love and of well-understood human interest. Not everybody can be a genius artist, but within the frame of position and profession everybody can contribute in a certain way to the enhancement of beauty, even if it were just by cleanliness and love of order. It has been tried many times to forbid the forester to do more than that. For example, G. Heyer wrote: "If a private person manages his forest in a park-like way and keeps all the picturesquely beautiful but economically mature trees in it, he cannot be prohibited from doing so, for he has to account for his property and its management only to himself. But it is different with federal forests. The revenues coming out of these serve to decrease the tax burdens of all the citizens, so even the poorest one is interested in the federal forests bringing high net profits, for then he will be required to pay less taxes."

However, elsewhere it has been pointed out with much empathy that especially the poorer people have a prior right that their enjoyment of the forest will not be detracted by financial narrow-mindedness. Among several others, Guse has repeatedly supported this point of view.

Once, when he was chief forest ranger in the city of Trier, while fighting adherents of the net profit theory, he considered the question of whether in management the state has to look only at financial points of view, and especially in favor of poor people; he felt that he had to answer this question in the negative. I would like now to quote part of his remarks:

"I would like to take the gentlemen for a walk through some parts of my districts. The day shift is over, the night shift begins—the mine workers return home on the miners' path through the oak and beech forests;—or walk on the same path for the descent. Look, how much these people enjoy inhaling the clear forest air! Nearly rapt, they look up to the high tops of the trees. Take a good look once more at these stands, they will not gladden the hearts of your children any more, for forest management of the future will not allow them to grow that magnificent any more. The industrial founder keeps his park—nobody can deny him that—but how about you? You will not even keep young beech stands, for softwoods will be more profitable. And furthermore, you will have to accept that you have to hurry home through fields without shade, and your children will not be able to play anymore in the nearby forests, for the property value will be so high that it would be ridiculous to continue to forest this place." . . . "Truly, it is not the poorer classes which will be put to a disadvantage by conservative forest management.—Wealthy people can travel and take impressions home with them, which will comfort them over the daily surroundings, but poor people are not in the position to do that. But if it is luxury to keep a high forest capital at a low net profit basis, a luxury the national economist cannot tolerate: Well, so we should be consistent as well in some other matters. Look at this railroad station, built in Norman style, for a huge sum of money, with comfortable waiting rooms for first and second class (for the rich people)! What wastefulness of public wealth! Is this profitable? Is this necessary? Over there the bridge, what an expensive decorative construction! And even more all these school buildings, parliament buildings etc. What a senseless wastefulness, in its whole much more than the one which lies in our long rotation periods. Why only limit oneself to the necessity? Either—or! Man has to be consistent, especially the mathematician."

Recently, Guse has confirmed his opinion. In a literary fight with Denzin over the permissible rotation period for pine stands on good soils he writes:

"Mr. Denzin looks a bit mockingly on my remarks about the greater beauty of forests managed in long rotation periods, on the 'idyll' as he says in which I portray forests managed according to my theories. He does not deny that forests gain more beauty with increasing age, but says that according to its main destination the state forest is an 'economic forest.' Who is going to deny that? But is a forest managed in long rotation periods not an economic forest? Does a forest cease to be an economic forest if it is managed in rotation periods longer than 100 years? And after all, is beauty not a criterion that our administration is supposed to consider? Hardly any railroad station on a minor track is built, hardly any public building, without this criterion being considered. Why should forestry alone be an exception? What gives more pleasure to the larger part of mankind, a stylistic building or the magnificence of the forests? Forests and lakes are the main attraction of the eastern parts of our country; are we supposed to reduce their beauty in order to increase the speculative land value, which nobody is going to pay to us anyway? According to Denzin the power of nature is visible in the rotation period with the highest mass production, 'which is still below the financial one.' I doubt that someone admires the **power** of nature more in a dense stand of thin sticks than below 200-year old oaks or in a stand with mast spruces and hardwood middle and ground layer. It is said that we are living in a materialistic era; nevertheless, the need for beauty is acknowledged in all the public institutions. The state is proud to be able to decorate its cities with memorials, buildings, etc., whose permanence is calculated to be many centuries; is a flourishing country not supposed to find its pride also in having a supply of magnificent mature forest stands?"

Whoever does not wish to share this ideal attitude of Guse should nevertheless try to avoid fiscal narrow-mindedness in questions of beauty, for taking care of beauty in the forest not only helps the owner with personal enjoyment, but is really worth it, in providing some concrete advantages.

To prove this in detail was the purpose of a longer discourse I published in the *Danckelmannsche Zeitung* [Danckelmann's Newspaper].

From that source I here indicate only the main points, because otherwise I would anticipate the applied part of this book.

1. The consideration of aesthetic points of view **protects against economic mistakes,** because along with the striving for beauty, which leads to perfection, one achieves the good and therefore the useful at the same time.

2. The eagerness of the forest civil servants to do their jobs is also dependent on the beauty of their districts.

This has already been pointed out several times by Pfeil in his discourse: "Knowledge by itself does not work, if there is a lack of love."

There is a certain kind of love for the forest, which should be instilled into the young forester, and Pfeil states that "it is not the kind of love which the timber merchant feels for a tree which is going to make him a huge profit, for this is similar to the butcher's love for a fat ox or a fat hog, and it will document itself only in cutting down the tree or slaughtering the animal. Nor is it the type of love which is rooted in vanity in order to be able to show off beautiful stands for whose existence the one who is showing may have done little or nothing. Even less is it the jealous love which desires to exclude all other people from using the forest in order to be the only ruler and to be able to treat and to use it arbitrarily, for this one roots always in condemnable egoism, even if only the manager puts himself at the position of the owner; but true love for the forest always goes along with love for people. It is the intimate interest in the prosperity of the trees and the forest, the striving to enhance it for sake of the trees themselves, and the willingness to make for it every personal sacrifice, to take over for it any effort and hardship."

With foresters who lack this love, even "the greatest knowledge" cannot substitute for this defect, according to Pfeil, for they will "not have their own eyes everywhere," and "another thing is that these dutiful indoor-foresters usually tend to bureaucratize the whole forestry."

Even today there are foresters who love the forest out of understanding its beauty and who know that the eagerness to do the job is partly based on this love; this is expressed very nicely by Forester Roedler:

"Do not say that all this is unnecessary, that it is useless nature-enthusiasm. Especially in the often hard everyday monotony of the service, the visiting and keeping hold of nature's beauty provides pleasant and often educating change and recreation. The thinking man, the true nature lover is pleasantly touched when he enters a forest district where he can recognize the enhancing and protecting hand of the keeper; the silent acknowledgement of the hiker, who enters the sacred cathedral of the German forest, is bound to be his. But not only this reward do I mean, no, also he himself will find high satisfaction and continuous stimulation in looking at his creations; plentiful pleasure is going to reward his efforts."

3. The **population's affection,** turned towards the forest for sake of its beauty, **is useful to the forest in numerous ways.**

It is hardly possible to express this better than Head Forester Neumann did, when he wrote: "The **beauty care** of the forests does not only satisfy an ideal mental need of humans, but also is a **means of protection,** for a well-kept, aesthetically beautiful forest wins the interest of all educated people, and just by itself, it is going to put itself under the protection of the majority of its visitors; this is going to contribute not insignificantly to the security of our forests. However, not many people subscribe to this attitude yet; officially it is not acknowledged at all, but it deserves attention in a full amount."

In addition I would like to mention that it is not only "educated people" who know to appreciate an "aesthetically beautiful forest." Also **for less educated or uneducated people, beauty possesses a privilege.** Not the uneducated human, but only the mob, spoiled by contact with half-education and brutalized, is going to desecrate the beautiful.

4. The enjoyment of the beauty of a nearby forest **is going to make the population settle down.**

This is something not only state foresters should consider for the reason of social politics, but also the urban and rural communities and private forest owners. Every landowner who wishes that his sons and grandsons are going to live on the soil like him, on which the fathers and grandfathers have worked and bustled around, has on the one side to bring up his children unpretentiously, and on the other he has to look after the profitability as well as the beauty of the property in order to make them love it. However, not everybody can possess forest property, but this is just the fortunate thing about the enjoyment of beauty, that in abandoning oneself into enjoyment and sharing the pleasure of the owner, one somehow begins to feel like a sort of co-owner. **"For he who enjoys, possesses, and can the owner do more than that?"** (Count von Moltke).

The owner, however, looks with increased awareness. Besides the pure aesthetic satisfaction, he is filled with another pleasant emotion, which is closely related to the first one. This is what I once called "the pride of forest property." In this connection I do not mean the legal term of property. Like everyone from Silesia talks about the beauty of **"our"** Sudeten Mountains, so every German regards himself as a co-owner of the German forest. That's why this point of view should be considered not only for the private forest, but also for the state forest, and especially, however, for the community forests, for every citizen, every member of the possessing community feels as a co-owner and enjoys this peculiar type of satisfaction, which in itself is not really an aesthetic one, but whose extent is determined mainly by the beautiful care of the forest.

Unfortunately, this charitable attitude still has some opponents. It is no longer denied that near populated cities, near spas, in hunting districts of high gentlemen or noblemen and in similar preferred forests, e.g., near well-visited tourist roads, something has to be done in favor of the beauty, but into remote large forests a forest aesthetician should not dare to invade, according to many people's opinion.

Remarks like that are hardly being found in literature anymore, but in reality you can learn that the vast majority of foresters still shy away from listening to aesthetic's voice also in the center of great forests. This is a consequence of the modest restraint taught to the forester, which, however, goes too far in this case.

First of all, the forester himself is "a human, so to say," and he looks at the forest every day and with understanding. The forester is dependent on the forest, and therefore his opinion is more valid than that of many tourists.

Then we should consider that the development of transportation routes has already made many forests accessible, although nobody would have predicted that. Because you cannot create nice trail layout, mature trees and some other things in a minute, like you can construct a villa, the forest has to be prepared for everything that may occur.

After all it would be quite a narrow-minded attitude if only what is often seen and seen by many people, should be shaped beautifully.

The human does not enjoy beauty with his senses, but with his intellect. We are delighted by the beautiful, if we know it exists, even when we do not see it.

When they supplied the tympanums of their temples with these everlasting exemplarily carved columns, the Greek masters turned the same attention to artistry to all the sides of the marble, even to the sides that would be remote from the spectator for all times, as they thought. Surely they must have been ruled by the thought **that not only the beauty we can see is delighting us, but also the beauty whose existence we know without seeing it. The same is true for our forest. Even somebody who never ever enters it must be caring for its existence and its beauty.**

One is tempted to regard this as self-understanding; but because this principle is violated so often, I would like to discuss this thought a bit more.

In enjoyment of the forest, things are related similarly to the enjoyment of horseback-riding: even if a horse is running very well and has the most pleasant movements, one will not enjoy riding it, if it has a severe blemish, as, e.g., lack of tail hair, even if one will not be able to see the fault while riding. Similarly, the forest owner will prefer to draw his revenues out of a beautiful, well-tended forest instead of out of a boring, unattractive forest.

In the Schlesischer Forstverein [Silesian Forest Organization], once the Koeniglich Saechsische [Royal Saxonian] Head Forester Schulze talked very clearly on the value of a beautiful forest property for a whole population, fortunately quite in contrast to people like G. Heyer. He pointed out that as proud as every Saxonian is to have the Sixtinian Madonna along with other art treasures in his home country, every Saxonian is just as proud to have with his state forests a valuable, beautiful and pleasant property, and it is commendable to consider the edifying, the beautiful and the pleasure of forest property along with the financial advantages.

3. History and Literature of Forest Aesthetics. Necessity to Study and Teach Forest Aesthetics as Special Branch of Forestry Science

The question of **whether it is necessary to teach and to study forest aesthetics as a special branch of forestry science** has been proposed several times; it has seldom been answered in the affirmative, but very often has been answered in the negative. Therefore the forest aesthetic literature has developed only very slowly. The first forest aesthetic writing was published in England in 1791 by Gilpin, canon in Salisbury, pastor in Boldre in Newforest. A German translation was published in Leipzig in 1800; it contains two rather impressive volumes with the title: *Bemerkungen ueber Waldscenen und Waldansichten und ihre malerischen Schoenheiten, von Scenen des Neuwaldes in Hampshire hergenommen* [Remarks on Forest Scenes and Views and Their Picturesque Beauties, Deducted from Scenes of the New Forest in Hampshire]. This book is very rich with fine and accurate observations and still valuable today.

By the way, it is curious that clergymen and other theologically educated people have often led the way in forestry sciences. Besides Gilpin, there is Vierenklee, the inventor of the increase formula named after him, Stahl, the founder of the first forestry journal, and Bechstein, the founder of the forest academy in Dreissigacker.

Gilpin's book seems to have been forgotten in Germany as well as in his home country.

It took a long time in Germany before there were some similar independent movements. In 1824, von der Borch published an estimable forestry essay in the *Sylvan,* and in 1830 he was looking for a publisher for *Aesthetics in the Forest,* which he wanted to edit decorated with pictures. Later on he complained about attacks he had to face due to this intention. His *Aesthetics* has not been published. In the meantime, landscape designers, foremost of whom is Prince Pueckler, botanists like Schleiden and other friends of the forest—I mention Masius and Rossmaessler as the most deserving ones—have strived to increase understanding for the beauty of the native nature of the forest. However, they were not capable of significantly influencing the current methods of forest management.

On a few pages of his *Forest Care,* G. Koenig condensed a surprising amount of noteworthy forest aesthetic hints. His remark, "A forest in its highest forestry perfection is also in its most beautiful state," demonstrates his deep understanding.

Burckhardt in his book *Sowing and Planting* has dedicated a remarkable chapter to "Forest Enhancement," and has published here as well as elsewhere many quite good observations and tips, but I cannot agree with all he says. In general he writes in regard to the beauty care of the forest: **"It is hardly possible to act after prescribed rules, but after that which the concept of forest beauty inspires."** This sentence will be valid in so far **that a purified concept of forest beauty in addition to a humble attitude certainly is the main ingredient in forestry art.** Such an unerring master-judgement ("Only the master is able to break the main form"), however, has been given as a natural talent to only very few people; instead of this gift, a regrettably uniform taste has settled with many otherwise well-talented professionals.

Gayer has pointed out several times that a sense of beauty channeled wrongly may even lead to economic mistakes, and there is hardly anybody who has never seen how some group of useful advance reproduction was destroyed, and with it a favorable species mixture, because the seed rows, the planting rows were supposed to stretch straight forward from one end of the plantation to the other without any interruption. "These are in my way," a forester once answered me when he cut down the few beech sprouts left over after a clear-cut in favor of the pine seed.—Furthermore, you might want to remember how often plants are wasted under high thickets where they surely will not have any growth space, just in order to make the plantation look pretty neat and complete. Certainly every administrator of a larger district has already complained more or less about this frequent violation, without thinking about maybe similar consequences of his own activity and for related causes; for example, it can be done easily while regulating paths and creeks and while marking compartment boundaries with a preference for the straight line and with regard to a nice image on the map.

But even for the well-talented and not in any way badly-influenced forester, it is important to know the rules of the art, as well as for everybody else who applies himself to an art.

The forester cannot, unlike an architect, a sculptor, or a painter, first make a sketch and a model and then still make corrections according to his wishes, or eliminate something, for his activity is not allowed to cost any money; quite the contrary, it is supposed to make as much money as possible.

After all, it is true that the healing hand of nature will balance many mistakes we have made; it will enable some beauty to unfold itself although we were standing in its way. But often we make it too hard for nature to heal. Many mistakes made in haste cannot be repaired within several generations.

If I were going to **explain this with examples,** I would be anticipating the remainder of this book; **but if there is one or another among the dear readers who is not going to drop this book without having corrected some previous attitude or without having received an inspiration for some rules, then hereby the proof of the usefulness and necessity of forest aesthetics as a special branch of knowledge will be practically given.**

The eight lines above have been transported from the first edition of my *Forest Aesthetics* into this place. In the meantime Wilbrand has repeatedly proved the necessity to study forest aesthetics, and not only occasionally and out of books, but **in lectures**

dedicated especially to this purpose. This last wish has echoed without response; likewise in literature, relatively little teaching material has been collected, and the best of that has been contributed by Wilbrand himself. According to my knowledge there were also essays on forest aesthetics by von Fischbach, von Guttenberg, Hempel, and Kraft, all of whom presented many good and new thoughts to their readers; forest organizations dedicated some hours to discussion on these subjects, and even the Prussian Upper Chamber debated a forest aesthetics request in the commission and in the plenum; furthermore, the author of these lines strove to supplement the existing work with additional contributions; but all this has remained just unfinished work, not at all to be compared with the enthusiasm which has to be admired in other areas of intellectual forestry work.

May the new edition of this book be successful in gaining some new and beneficial friends for this subject!

Chapter Two.

The Causes of Delight in Beauty.

1. Man's Intrinsic Ability to Get Pleasure from the Beautiful.

"We enclose beauty in beautiful things as a remaining intrinsic quality; we claim that they themselves are and remain beautiful, even if we did not recognize them, did not appreciate them. It remains a beautiful statue, even if it were sunk deep down into the sea. Both the beautiful and beauty would keep quite the same value, even if it were not as rare among living beings as it is. Beautiful things can be significantly related to us, but they do not get beautiful just because of this relation to us."

If we accept this sentence of Krause, it will raise two more questions:

1. Which are the properties of human nature that enable us to get pleasure from beauty?

2. Which are the properties that are intrinsic to the things which correspond to our sense of beauty?

Let's take a look at the first question now. Darwin transfers the location of our aesthetic judgement only into the senses. In the second volume of his book about the origin of humans he writes: "The senses of humans and of inferior animals seem to be such that they enjoy brilliant colors, certain shapes as well as harmonious and rhythmical sounds and call these beautiful; **however why this is so, we do not know.** Certainly it is not true that in the human mind there is a certain standard for physical beauty. However, it is possible that in the course of time certain tastes have become inheritable, **although there is no evidence in favor of this opinion.**"

Although Darwin is rather modest here (visible in the lines emphasized by me), some lesser minds have not always been like that. For example Rossmaessler, who is otherwise very qualified in the field of forest aesthetics, writes: "Possibly the opinion could be suspected here, that nature has brought itself along to the formed taste of mankind, which opinion would be coherent with the one that puts man into the center of creation and subordinates everything to his interests. Especially those have this presumptuous judgement who call themselves the most humble ones, and they are easy to contradict. It is not the tree and the plant kingdom that have been set up according to human taste, but human taste has formed itself little by little because of their presence. The old German building style, with its overly rich decor of leaves, pointed arches, and roses, points as much to our German forest, as the ancient Greek style of columns does to the simple beautiful palm of the South."

In spite of Rossmaessler's attack, I cannot believe that there are or have been people who had the opinion that nature has brought itself to the formed taste of humankind. He has flailed into the air and has not violated anybody. Rossmaessler's remark, however, is brilliant insofar as it contains a half-truth; for we have to agree with him insofar that long habit makes the outside world familiar to us, makes us love it, and finally makes us think about things as beautiful, which are relatively unsightly in comparison to more beautiful things. Out of this taste, which has been determined by the surrounding nature and which has been inherited and consolidated from generation to generation, the often obvious correspondences of nature's appearance with the results of human crafts and arts can be explained. In addition to Rossmaessler's example, according to some aesthetic philosophers' opinion, the dominance of green in Persian rugs is due to the rich vegetation of the Persian mountains and the fiery colors of woven fabrics of Africa are due to glowing illuminations of the desert.

But after all, Rossmaessler's attempt of explanation remains totally insufficient. We can never admit that we like nature **solely** for this **single** reason, that is because we have got used to it; otherwise our aesthetic liking of its treasures would have to go along with the knowing, with the habit, which however is not true at all. Furthermore, I want to disregard the fact that the first impression may have its attractions, too, but I would like to point out that to hardly anybody the useful tuber of the potato will appear more beautiful than the flower of a rose, however much he may study it.

But if the liking of nature is not a matter of pure habit, then what is its foundation?

Maybe coincidence? It must have been very intricate coincidences which would arrange the atoms out of which the outside-world is made, and those other ones which build up our bodies in a series of changes, each time in such a way that, an aesthetic liking of the former ones would develop into a matter of consciousness in the latter ones. I do not doubt at all that we do not lack detailed attempts of explanation, too, in this direction, or soon are going not to lack them, for some have tried to explain even the moral properties of humans on the same foundation; but I think the dear readers will agree with Theodor Hartig, who stayed until death with the opinion that he could not imagine an elaborate building without an architect.

Here I quote the remarkable confession of that outstanding researcher: "The empire of insects has plenty of cases in which the most different bodies of nature, here plants and animals, stand in interdependent relationships to each other according to nature's law. These are quite incompatible with the modern point of view, according to which even the thought itself is thought to be a product of material functions, of a turnover in the brain's substance. Mechanisms as well as organisms need a special guiding force whenever they are working, that means when their movements are bound to serve a given purpose or a certain goal, which in regard to the inanimate machine is called helmsman, foreman, etc., in regard to organisms is called vitality; a force that cannot be within a material, because it is going to move and to change different materials in the same way and the same materials in a different way."

There is a special difficulty in explaining the delight in beauty in a material way due to the fact that the fight for survival is by no means always made easier by those changes in shape and color whose appearance we think of as beautiful.

If it is not pure habit and not coincidence that the surrounding nature appears to us as beautiful, then the materialistic theory is going to fail even more in regard to art's beauty.

Our pleasure in beauty, insofar that it is a purely aesthetic one, is free of any personal interest. Merely the contemplation or cognition of beautiful things allows us pleasure. Out of the struggle for survival this appearance cannot be explained; quite the contrary, if **only** the struggle for survival would rule the organic world, then the impracticable hindrance of aesthetic points of view would have had to be discarded long ago.

Some people, however, have been highly successful in this discarding. But how do we judge these people? We do not admire them as if they were advanced, but we condemn them, and in that we prove that we possess an **aesthetic conscience,** which demands consideration of the world of beauty just as the moral conscience does in relation to right and wrong. It can often be seen that people dispute and argue as heavily or even more intensely on the question of whether something is beautiful or not as on the question of whether something is good or evil. Everybody tries to convince the other one with **arguments of reason.** The often quoted saying *de gustibus non est disputandum* [it cannot be quarreled about taste] ends the fight only formally; but each one still thinks that the other one should have agreed with him, and it would not only be unreasonable but even wrong to hold on to an opposite opinion.

This, however, so to speak, is a matter of emotions. Whoever wants to see proof of the correctness of this impression should read Oerstedt, who explains in witty talks about the circle, about the fountain and the reasons for the pleasure which is created by sounds, that our faculty of sensation obeys **laws of reason** and that the rational laws which rule us are the same ones which rule as well elsewhere in the universe.

Recently Jungmann has expressed himself in quite a similar way, following after Plato:

"The beauty of things is their actual correspondence with the rational mind, insofar as they are suitable to be objects of pleasure to the rational mind by this correspondence."

Our ability to enjoy beauty accordingly stems from our reason. It is more the task of religion than of aesthetics to answer the deeper question of the origin of our reason.

2. The Properties which are Attached to Things and Make up their Beauty; Teaching of Adornment

Now we ask: Which properties of the things are these which can be objects of pleasure to the rational mind in actual correspondence with the rational mind?

As examples for all that is beautiful, we visualize Cologne Cathedral, a butterfly, the Ninth Symphony of Beethoven, a forget-me-not, the marchpast of a cuirassier squadron, Schiller's "Wallenstein," a sunrise, an oak, Shakespeare's *"Romeo and Juliet,"* the Prince Bismarck, an orange, and we ask ourselves: What do all these things have in common?

First off, **in all of them a certain idea manifests itself.** The embodied ideas are quite different from each other, but each one finds a very perfect and perceptible realization in the listed beautiful things. If anybody does not feel like this, then I would have chosen my examples badly. To me, the sublime building in Cologne is the ideal of a cathedral, the forget-me-not at the creek is the ideal of a spring flower, etc.

Goethe makes Eugenie speak: "What is the appearance which lacks essence? Would essence exist if it did not appear?" So he thinks of appearance without essence as being invalid, appearance has to correspond with the essence of a thing. In this sense, beauty is based on **truth.**

Therefore we demand from a church that it make us recognize even from far away its relation to celestial things by a soaring tower or by a dome imitating the firmament. Furthermore, it is supposed to make us recognize its material (stone, brick or wood) and its era of emergence by its style of construction. We even wish that it expressed clearly the special excellencies of its architect. In short, works of art and works of humans in general are supposed to have **one distinct** style.

The buildings of Schinkel: they are beautiful after all, but how much more would we owe this master, had the right material been at his command, or had he decided **to develop his style** according to the **existing** material.

Regarding the formations of nature, it is not possible to talk about style; however, our taste has similar claims on them. We like mountains the better, the more they reveal a plutonic or neptunic origin; the husky cry of the seagull pleases us only because it corresponds with the restless hustle and bustle of this land-avoiding bird; the characteristic mark on the wings is a main decor of the genus of the *Noctuiden* with a large number of species.

If a thing corresponds deeply with the ideas we have of its genre, **if its appearance is a typical one,** then the expert may greatly like it; nevertheless it is not really beautiful. This is the type of attitude with which the physician can regard objects, not only with interest, but with pleasure, which are repulsively ugly to the layperson.

A seeming exception proves the rule, inasmuch as some arts, which we are accustomed to considering as the source of highest beauty, aim at creation of a beautiful **appearance**; however, particularly the artistic performance is reproached if it cheats us. For this reason wax figures are not pleasant, and a painter should never go so far in forming some single objects in the foreground that they appear nearly plastic. This is quite natural, for nobody likes to be cheated, except for a harmless joke as happens in the well-known competition between Zeuxis and Parrhasius on part of the latter. The example of the physician who eventually might get pleasure from diseases that horrify us, however, proves that it is not unimportant which **idea makes its appearance**. Ideas are not equivalent and much is dependent on the type of idea that appears. For example, a cross-spider may be the most beautiful of spiders, but it is far behind in comparison to a butterfly's beauty. The butterfly, however, is far behind in comparison to higher beings. It pleases us inasmuch as it is really perfect. The human body, however, has such an abundance of beauty, that even after bad mutilation its ideal is still able to arouse enthusiasm, like the torso proves, which aroused Winckelmann's artistic taste.

Besides the many other circumstances according to which we determine the value of ideas, especially the moral requirements of our conscience weigh heavily. On this, the close **relation between the good and the beautiful** is based. The use of language alone shows how closely related both terms are, as often the words beautiful and good are used in equal status. We speak of a **beautiful** deed, when we really mean a good one; the Greeks liked to use the words χαλος and ἀγανος in combination with each other: they called the honest man χαλος χἀγανος, for it would have appeared to be a contradiction to them, if the honest man had not been beautiful as well, or the beautiful one had not been honest as well.

In this respect, Socrates says: "You are truly beautiful, Theaetet, and not ugly at all, unlike what Theodorus said about you; for someone who knows how to talk intelligibly is beautiful and right."

Shakespeare, the philosophical poet, expresses himself in a similar way, when he makes the Duke of Brabantia speak in regard to Othello:

"If one has to recognize virtue as beautiful,
then you are not allowed to call your son-in-law ugly."

Klopstock wrote in a similar vein:

"Oh virtue, I called, virtue, how beautiful you are!
What divine masterpieces are souls, which elevate themselves
up to you!"

I have to cope with an objection that suggests itself. One can point out to me numerous works of art whose morals are rather doubtful, but which after all are generally regarded as beautiful, e.g., Apollo skinning Marsyas or the swan and Leda. I respond: such works of art are comparable with Schinklel's buildings. Just as it does harm to these, that their walls, their timberwork and their decorations inside are not what they want to appear outside, the painter's and the sculptor's works mentioned above are **far below those which express an uncontestable idea with the same skill.** Only the most extraordinary mastery of form that the artist reveals can make us accept the lack of inner content, but who would tolerate, in the form of a plaster cast or a common oil-print, a mediocre copy of these works in his room!

If, however, the idea of the good is the highest, and whose appearance, therefore, will grant the highest aesthetic satisfaction, so the idea of the good in common life, that means the **useful**, will have to be put at its side as being closely related.

When we mentioned above that the outer form has to correspond with the material, we already touched the term of **usefulness**, and now again our path leads up to it. We think of foolish, useless acting as wrong; that is why we are hurt by it, similar to a moral lack, while clearly recognized usefulness is pleasant to us. The human, a rational being, can only be satisfied by the reasonable—(and this is the useful)—and this principle of course rules our aesthetic needs as well. Here are some examples that might take their places in order to clarify this: Greek columns of different grades are only as close to each other as it is necessary to carry safely the burden placed above. The stronger Doric ones therefore have larger distances between each other than the Corinthian ones. In lightweighted timbering we like slim columns which are distant from each other, and in iron constructions the eye demands the widest extensions with the greatest slimness of the supporting beam. We like the light construction of a winter garden, made out of much glass and some iron; with a bridgehead we like the solid wall with narrow windows and battlement circle. Furthermore, we like a building only if it has not been erected without a purpose, may it even be perfect. For this reason the small temples and similar things which were used abundantly in former times for decoration of parks had to be rejected, not to mention those bridges, many of which were drawn in a high bow across a lawn depression.

We might as well have chosen the opposite way and might have started with the call for usefulness, reaching the call for the good, as proves the remark of a famous aesthetician (Fechner). He says: "the best taste is the one with which comes up as a whole the best for mankind; the better for mankind, however, is what is more favorable to its timely and probably eternal welfare."

Our evaluation of the content of ideas not only depends on the terms of good or evil, of useful or useless. We set everything human above the animal, animal life above plant life, the world of plants far above the world of inanimate rocks. In general this is in accordance with the ladder of our aesthetic evaluation; but it does not exclude that the formations and appearances of inorganic nature may often be regarded as marvelously beautiful, too, for they substitute by size what they lack in liveliness, and the pious poet, who wrote the *Book of Job*, has not remained the only one who heard the Lord's voice out of the tempest.

I am quite aware that aesthetics, according to a certain more strict limitation of the term, has nothing to do with good and evil, with useful and useless, but is limited to what our senses transmit to us. Or, in other words:

For a stricter philosophical attitude, things are only objects of aesthetic pleasure, inasmuch as our judgement of taste on them is not influenced by any ulterior motives concerning their purpose and their nature, but only based on **their way of appearance**. This attitude would not help us with our goal, for **a totally uninfluenced judgement of taste can never exist in reality**. As we cannot change the way we are, we cannot lay down our experiences and our knowledge.

Another example may clarify this:

I invite the dear reader, who has followed me so far into dry theory, now to go with me into the green spaces for recreation. We go together into the forest. At about dusk we see on the first clearing something reddish. Is it a dying fern, is it a roedeer? We cannot distinguish it and do not pay attention to it anymore. Suddenly the red point moves, so then it really is a roedeer, and even though it now again does not move anymore, and even though we now can see less of it than before due to the increasing darkness, now the vanishing red point is sufficient to enhance the whole image of the forest; for now we do not see but we know that the forest is living.

Up to now we have learned about the correspondence of appearance with the basic idea that it is a requirement of beauty, but with this the nature of beauty has not yet been sufficiently explained. Besides this, the nature of beauty is based on **harmony**, or, to use a technical term, on **unity in variety**.

Nothing we perceive exists just for itself, but it exists only as a part of the world, and only this, the cosmos, is actually the beautiful; to the limited human mind even parts appear as the whole, as soon as it can see a clear boundary, and as soon as the smaller parts do not lack a common character. The pleasantness of these wholes can be supported from two directions. On the one hand the relation to unity can become evident (think for example of an all-over, one-colored wallpaper); on the other hand the parts can become evident. The most efficient relation is given **if along with strong division unity is kept after all.** This is the case, if each part stays incomplete in a way, that one needs the other for its completion. A large building in Renaissance style which is clearly divided into basement, stories and roof, reveals in its parts much more diversity than is usually shown in a huge complex of several modern city houses united to a part of town; with the former, however, no part can exist without the other one, while each one of the tenement houses can quite well be fancied without the other.

The size of the parts has to be in reasonable relation to their values. One probably is going to give windows of the same size to equal rooms in one and the same floor, but the main floor of a country home will demand larger windows than the basement and the upper floor.

Even when dividing into **unequal** parts, the **relation to unity** can be maintained very definitely and in a beautiful way, if one and the same proportion connects the parts among themselves and with the whole. This is enabled by the **golden section**. Here the smaller part relates to the larger, as the larger relates to the whole. However often one is going to divide further in the same proportion, there will never be insoluble confusion due to the always maintained law. Measurements have proved that indeed the forms which we recognize as the most beautiful ones are arranged according to this proportion; as e.g. the human body and the all time masterpieces of architecture, as well as plays which tie the knot in three acts and solve it in two.

The teaching of the golden section has been rather neglected in forest mathematics up to now; it has been overgrown by the calculation of compound interest. Therefore it will not be superfluous that now we are going to recapitulate together briefly what we learned in high school:

If a line a-b is to be divided according to the golden section, a perpendicular line is to be erected on one of the ends (a c) with the length ab/2; after that the hypotenuse between b and c is to be drawn, and from this a part with the length of bd = ab/2 is to be cut off. There remains a part cd of the hypotenuse which of course is larger than ab/2. We call this length M and we are going to cut this length M from ab. After doing so there remains the rest of ab, which is smaller than ab/2, and which we now call m, and there is the proportion ab : M = M : m.

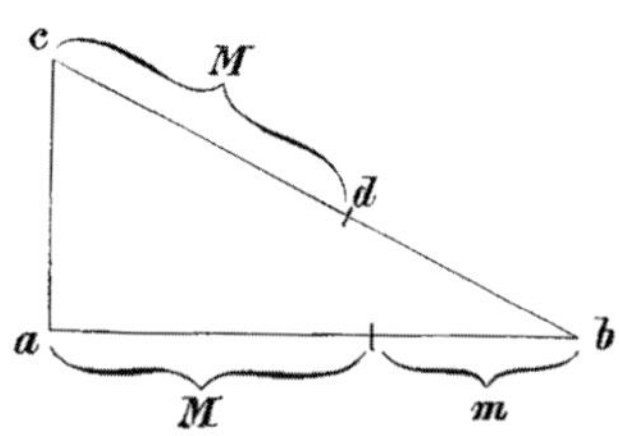

Fig 1.

I do not want to detract from anyone's pleasure in finding the proof again by his own power, and I have the same attitude towards the arithmetic theorem according to which the first back-part relates itself to the second one as the difference of the parts of the first proportion relates to the difference of the parts of the second proportion. For this reason we can state:

M : m = (ab-M) : (M-m). If we call M-m = m_1, then we can write M : m = m : m_1; this can be continued into the infinite with m : m_1 = m_1 : m_2, etc.

This proportion cannot be expressed in whole numbers, but one can approximate it with numbers from the sequence 1, 3, 5, 8, 13, 21, 34, etc.; each one of which is the sum of the previous two. The larger numbers offer higher accuracy; 8 : 13 = 13 : 21 is closer to a correct proportion and to the golden section than 3 : 5 = 5 : 8.

In respect to the human body, the lower part up to the waist corresponds with the larger section (M), and the upper part with the smaller section (m). We find this part (m) again in the section from the waist to the knee, and therefore there remains $M\text{-}m = m_1$ from the knee to the sole. This length (m) measured from the waist up reaches up to the larynx, so there remains for the head $m - m_1 = m_2$. Like that, one proportion dominates the whole human body. It dominates, too, the wavy lines which outline it.

Wavy lines are especially pleasant if their height relates itself to their length like m : (M+m) or like M : (M+m), and if also the edge of the roof of the wave relates itself to the rising part like the proportion m : M.

However, much is dependent on the **arrangement of the harmonious sections.** With equal parts, **symmetrical arrangement** makes the survey easier and enables parts of **the same kind** to unite.

Color and adornment may help additionally, and **homogeneity of movement** has a unified connecting effect; think for example of a flying group of doves or starlings. Finally, the **term of usefulness** has to be considered again now, because often one and the same purpose creates the most efficient relations between things that are otherwise not connected with each other. How different in nature and form are the things that decorate a cultivated dinner table, and how beautifully unified are all of them because of the common purpose they are bound for!

But of how much use could beauty be to us if we did not understand it! The beautiful things comply with our need for understanding in a universally intelligible language, the **language of adornment.**

According to Selenka, who dedicated a detailed examination to the adornment of humans, **"adornment is nothing else than a natural language, universally intelligible, suitable to tell figuratively our advantages to our neighbor."** Human adornment is divided into groups according to Selenka, namely

1. **Ring adornment.** As such it is "proportional adornment," bound to mark the body's natural division, or bound to serve as "symbol of fullness of life" in provoking a swelling of muscular flesh (for example the arm-ring worn by ancient Germans on the upper arm).

2. **Hanging adornment,** which forms perpendicular or falling lines on or beside the body (for example the diadem of Priamos, or the pendent ends of the officer's sash). It is effective by contrast in "visually emphasizing the wavy and changing outlines of the organic form in its lively gracefulness." Hanging decor is a hindrance to quick movements of the wearer and therefore indicates dignity.

3. **Direction adornment** (e.g., helmet decoration made out of a horse's mane) is supposed to mark its wearer as highly mobile.

4. **Attached adornment** is bound to indicate the wearer as a respectable personality (e.g., grenadier's hair cap, train of a dress at court).

5. **Local color adornment,** (e.g., flowers in the hair, a precious stone in a ring) is supposed to attract attention to certain advantages of the male or female owner, e.g., beautiful hair, a beautiful hand.

6. **Clothes adornment:** The stiff neck collar of the soldier is "a symbol of the stiff posture and of the disciplinary obedience in general," the soft lying collar and the jacket with reverse collar are supposed "to indicate the not-buttoned-up, but open-hearted attitude." The simple street clothes of today indicate that their wearer is living in the era of business and trade.

I leave to Selenka the responsibility for these differentiations, although I myself would have differentiated some things in a different way, e.g., I would list direction adornment as a subdivision of attached adornment; but anyway his work has merit, it sharpens our judgement, and is useful not only for the appreciation of human adornment but rather useful in general. Some examples may show this:

1. Ring adornment: Nearly every table lamp displays it, as well as every blade of grass (with the single exception of *Molinia,* which is characterized by the lack of nodes). At buildings the ledges are supposed to mark on the outside the division of the stories.

2. Hanging adornment: The pieces of flesh at the turkey's head.

3. Direction adornment: The smoke plume of the running train, the hanging birch in the wind.

4. Attached adornment: Tail and feather comb of the peacock.

5. Local color decor: Colors of flowers, the blue upper wing feathers of the jay.

6. Clothes adornment: The stiff garment of the evergreen conifers.

That adornment has only a minor importance may be proved by an example: The fat hand of a lazy upstart will appear the more disgusting the more rings, or the larger precious stones are put onto the fingers; however, the strong hand of a loyal worker is beautiful even without ring decor; only the wedding ring will fit here and maybe the signet ring. A well-kept hand of intelligent people who talk gracefully not only in words, but also in gestures and sweeps of the hand, may bear a third ring.

Certainly everybody will beautify himself by a friendly expression, but adornments should be used carefully.

This will be valid, too, for the forest. It will always be quite safe to make the desolate pine heathland more friendly by using several birches. However, it is a courageous attempt to decorate it with *Abies nordmanniana,* and it will easily fail. But I should not anticipate. The forest aesthetic useful application of the teaching of adornment is going to be discussed only in the second section of the second part [of this book].

3. The Pleasure in Beauty Can be Increased in Various Ways; It Needs to be Protected Against Disturbance

The more beautiful a thing is, the more **our pleasure will grow in getting a precise knowledge of this thing.** I would like to remind the reader that we hardly ever are able to appreciate a good painting at the first sight or a high-quality work of music at first hearing in their whole value; I would like to remind also of the point of view of the scientist, who, familiar with the structure of insects by study, recognizes with the caterpillar the subdivision and with the millipede the order and does not dispute a certain beauty for both animals. It is said very often that the going into detail required by study "destroys all the poetry," and laziness takes this assertion as a welcome excuse not to study seriously botany or zoology at all. However, already Goethe and Humboldt were forced to cope with this mindset. The latter describes their attitude as being based on "limited intelligence and a certain sentimental dullness of the mind," but thinks of the question as being important enough to dedicate to it several pages of his *Cosmos* [handbook]. He admits, however, that staying with details does not inspire the mind ("and this is fortunate concerning the safe success of the work"), but at the same time, he emphasizes that one is going to have much more pleasure, and will be "led to greater views" after knowing details, than in the state of ignorance.

The gentle reader may want to look up all this later on in the Cosmos, but now hear from me, that we do not only take things the way they are, but add to them from our experience and enrich them by subconsciously connecting them with **idea-associations.** Such associations may be of various types. The most important ones are those which stem from a **relationship of our intellectual conditions** with the relations of the material world. In this way the bright sunny landscape is cheerful to us, the *Salix babylonica* is mourning to us, the dark cypress is related to our seriousness.

There is nothing arbitrary in this, but in those other associations, which Oersted calls "beauty by imputation." These are not ineffective either. So, although unjustified, nevertheless quite significantly, the lion is decorated with magnanimity imputed to it, the dove with its imputed gentleness and the violet with modesty.

Furthermore, the **ideas of the good and of the useful** are very important, again in an indirect way, and this not only depends on what a thing is now or will be in the future, but nearly more on what it has been in former times; even more effective, however, are the associations coming from **purely personal interests:** the thought of advantages one has enjoyed or is expecting. These are quite justified; nevertheless, every now and then some regrettable aberrations of taste spring from them.

Once I knew a farmer who thought a fat pig and a well-worked beet-field not only to be beautiful—and we ought to agree with it up to a point—but by far the most beautiful things in the world.

Hardly any other field of aesthetics is as controversial as the one just mentioned above. According to some authors, the idea-associations are most important for the nature of beauty, while according to others, they are not important at all. The following example may clarify the point of view taken here: Schiller's "*Wilhelm Tell*" pleases in general mostly because of the patriotism it expresses, and in this respect it is the idea of the good on which our pleasure is based. If, however, a Swiss man likes the play particularly because it glorifies his home-country, or a Swabian because the poet was not only a fellow-countryman but as well a very honorable person, then these are idea-associations, which in reality do not have anything to do with the true nature of the work of art. It is not seldom that such accompanying associations of thought determine nearly by themselves our judgement of taste.

Looking back on the things said above and considering that we get to know things better from year to year, and that they are enriched daily by connecting more of our thoughts to them, we come to the cognition that the surrounding world would have to seem more beautiful to us each year. Experience confirms this assumption, too, for the matured ages have far more sense for beauty than childhood and early youth. However, there are some exceptions, and these are caused especially by our **need for change of impressions.** To the mountain dweller, his familiar mountains become indifferent at last, and only after he has been away for a longer time, will he be able to enjoy fully their sight, but then certainly he will be able to appreciate their value much more than the dweller of the plain who travels to the mountains for the first time.

Let us now ask which **relations make us pay attention and retain our interest in things the longest;** we recognize such relations first as a real change in the things themselves, and second as an arrangement such that the things are not immediately visible in their whole, but **appear different from different points of view.** For the first one, we have an important example in our vegetation, whose nearly daily changing apparel compensates us more than enough for the splendor of tropical forests. For proof of the second one I remind the reader of medieval family mansions full of nooks and crannies which one can never see enough of.

Unfortunately our inability to stay continuously with the same impressions is often only the smallest hindrance in the pleasure of beauty, for this pleasure is dependent particularly on our **physical and mental well-being.**

Every pleasure is incompatible with serious worries, and temporary worries will decrease it more or less; likewise, we are so dependent on our body that its illness or even fatigue will dull us nearly as much as mental illness or mental weariness.

It seems to be unnecessary to discuss all this in detail, but I have to mention that one **possibly can be sensitive only to certain types of beauty, however doubly sensitive to these.** So a sad person may want to listen to serious music only, but may enjoy it even more in sadness than in an indifferent mood. Someone who needs physical rest may like to visit a garden with a distinct boundary—maybe brightened by a fountain—when nature's splendid unfolding in the high mountains would irritate him. A taste influenced in that way appears not only with the individual and temporarily with individuals, but the phenomenon appears more or less in the history of nations. So it is deeply rooted in religious and social conditions when the creations of art devote themselves now to the purely beautiful, now to the sublime, to the romantic or to the graceful.

I just used expressions I had not explained before. All of them name the so-called **modifications of the beautiful.** Out of the great number of these I chose to mention briefly those which I think are the most important ones for us.

The form of beauty which pleases by most perfect harmony of the parts in absence of sharp contrasts is called the **purely beautiful**, while the **picturesque** is based merely on the existence of such contrasts. Beauty in movement is called **gracefulness.** The beauty of such appearances which make us realize our smallness by their greatness or splendor rises to the **sublime.** The fall of the sublime we call **tragic.**

Particularly important to the forester is **humor** as a comforter in the numerous adversities of existence. The most marvelous flower of this genre of the beautiful in the field of forestry was given to us by O. von Riesenthal, whose "*Bilder aus der Tuchler Haide*" [Scenes out of the Tuchler Heathland] are really classic. Here the remark may be added that even the **ugly** still has some beauty. "Everything that is not completely beautiful is ugly at the same time, and everything ugly is beautiful from its reverse at the same time." Therefore one is supposed to learn to look at everything, and to present everything from its best side.

SECTION B. The Beauty of Nature.

Chapter One.

Preliminary Remarks on the Relation Between Natural Beauty and Artistic Beauty

Natural beauty is particularly important for us foresters who live more outdoors than many other professions and have to guard and to care for a rich treasure of natural beauty.

While the ordinary person regards it as self-evident that nature is beautiful, while springtime has filled poets with enthusiasm nearly as often as love, philosophers are not aware of whether and how much they are allowed to think of nature as being beautiful.

For example, Hegel (in this judgement I follow Zeising) has never known the existence of the beautiful within nature; to him only the artistically beautiful was valid as "the beautiful." Even Schiller, who proved his sense for nature many times, contradicts his better self as soon he begins to philosophize. In his essay "on the sublime" he presumes to claim:

"Now nature itself sets up objects in great quantity on which the sense for the beautiful and the sublime might be practiced; but here as well as in other cases, the human is served better by the second hand than by the first, and likes better to receive a prepared and selected material from art than to scoop arduously and miserably at nature's impure spring."

Hermann has discussed these questions more distinguishedly than others when saying:

"So the whole content of art is, so to speak, indicated and preformed in nature. Nature has, so to speak, wanted to create all its real things in the way they are presented or showed to us by art; as it were, there are these aesthetic basic ideas of nature itself which make up the content or the essence of works of art. So after all there is always something true and objective included in works of art, and finally it is only therein that the main value or the general importance of art lies.

The individual thing in nature is, on the average, always less perfect in regard to its appearance than a work of art, and it is just this imperfection which brings up in us the need for formation of art. In this respect, all art is at the same time a critique or condemnation of nature in its given individual things or appearances. Art would not arise in us if nature itself were perfectly beautiful and aesthetically satisfying. Therefore art, after all, partly recognizes nature, partly criticizes and improves it at the same time. But it is not to be assumed that nature in general or in itself is imperfect or has failed its real meant-to-be goals or has reached them somehow in an incorrect way. On the other hand, even a work of art, most perfect in itself, would have to appear to us somehow imperfect, unsatisfying or incapable of surviving, if it were a truly real or living thing. So, both nature and art complement each other and have to be measured with totally different and independent scales. The scale for judgement of art cannot be deduced or taken from individual natural things themselves. From the artistic point of view these natural things will always be somehow inferior or imperfect ones. Inasmuch as nature itself seems to carry the property of an artistic whole or of a well-ordered totality, it is not its individual things or appearances as such, but only its institution or its idea as a whole which seems to be able to be compared with the character of a work of art; and therefore, after all, the importance or the truth of art is the fact that it is for us a portrayal or appearance of the ordering thought of unity of the world's institution or of the real things' entirety. What art condemns or beyond where it goes is the empirically given detail as such in all the clumsiness and heaviness of its composed qualities; however, what it recognizes in nature or it expresses in itself are the organic thoughts and laws of life of the institution of all the real things. Therefore, the individual thing in nature is valuable for art everywhere only in this respect, that it contains at the same time the reference to some general natural law;

and the general task and the character of art lies in the individual thing's purification to its pure and higher value."

Hermann's assertion that "the individual thing in nature is, on the average, always less perfect in respect to its appearance" than the work of art, appears to me as unjust as the reproach further on that the empirically given details as such suffer from "clumsiness and heaviness of their composed qualities."

This opinion is rooted in the perception, quite correct in itself, that the artist can never simply imitate nature. While the artist redesigns for the specialties of his art the material nature brings to him, it comes with the vanity which exists even in great men that he thinks of his own work as more perfect than the original. But what would be said if the sculptor would try to raise his work above that of the poet, only because Lessing proved in Laokoon that the sculptor can use the material Homer brought to him partly not at all, partly only after complete redesigning.

By the way, Hermann does not miss that also a work of art, "if it were a truly real or living thing, would have to appear to us imperfect in some respect." He just refrains from deriving the corresponding conclusions out of this fact.

Unfortunately many humans disregard beauty in nature, even without the blame of philosophical teachers. One root of this evil lies in the unblessed habit of us common people to think little of things, which are granted to us without special cost or effort. How much is the urban citizen going to pay for a miserable, small bunch of flowers, how happy is he when he keeps a few ferns alive in his backyard, and what have we got in comparison? Only a little thinking is necessary in order to calculate that nature's treasures held by our forests exceed immeasurably the value of all collections of art solely by the value of their beauty, and in these forests we are the directors of the museum!

Ignorance, too, often is to blame when beauty in nature is disregarded. Even Jungmann writes without guessing how much he embarrasses himself in this: "What, however, concerns the pebble and the piece of wood, on which one cannot discover beauty; after all we limited humans, bound to senses, do not say that we **find** beauty in every thing, but that every thing is beautiful, that is, that every thing possesses inner goodness and therefore will be a reason for joy to the reasonable mind **who recognizes it clearly.**"

The correctness of the last sentence I experienced myself when 32 years ago R. Hartig and Remele demonstrated to me the "inner goodness" of the piece of wood and of rocks so clearly that I can hardly imagine how someone may not think of such things as beautiful. In the field of zoology, Altum opened my eyes, and this in such a way that I ended up with an actual appreciation of a work of art when I had started to understand nature. Traveling to Eberswalde, I looked at the famous group of Kiss, presenting Saint George fighting with the dragon, in Berlin. This work of art left me cold. I did not understand why the knight would not use the sword in fighting. Traveling home, however, I looked at the group with different eyes. Now to me the dragon was living, after Altum had disclosed to his audience an understanding for the weapons of animals. Then it was clear to me with one look, that the knight would not be able to harm this immensely armed monster by using a sword, that only the help of that higher power, which has the banner with the cross as its symbol, could lead him to victory.

Understanding of art and understanding of nature complement each other in the most beautiful way, as can be seen in this example.

Unfortunately I can dedicate only a few chapters to the contemplation of beauty in nature, while I pick up only individual pictures out of the overwhelming wealth of the subject, because otherwise this section alone would rise to the quantity of a whole volume. But I hope that the few things I can provide are going to be sufficient to increase love for the forest along with understanding of beauty in nature.

I was not able to dedicate special chapters to the aesthetic appreciation of rewarding fields, e.g., forest wildlife. I hope some more competent author is going to close this gap in forest literature very soon!

Chapter Two.

Theory of Color in the Landscape

The teaching of color in the landscape is a complicated subject, with which one can hardly straighten things out with oneself without having read one book or the other.

In advance I need to make some remarks about **seeing in general.**

If we look at a thing, a very small part of its surface is reflected on our eyes' retinas (e.g., with a deer it is not much more than some of the surface of those hairs which are directed towards us), and we do not **see** but **know,** due to endlessly repeated perceptions, that the existence of a thing in the direction of the look corresponds with these visual impressions. One learns best how much of seeing is a matter of training during the time when the first beard hairs become victims of the razor. How do we first need to learn then what we are going to use later on subconsciously with every movement in front of the mirror, that we have to convert every picture seen in the mirror before it will correspond with reality. In the same way we had to learn in early childhood to estimate correctly the size, the distance and the mutual position of things. However, we do not remember these studies any more nowadays, when, quite the contrary, we are not able any more to realize that, with a roebuck there is also a body going along with the surface directed towards us, the body of an animal. Everything else we know of this animal, the delicateness of its movements, its cautiousness, its character of a game animal, all this is brought to our awareness so clearly along with this one look, that we cannot overlook it at all (as linguistic usage puts it very strikingly). **So seeing is more of an art than we usually realize!**

The **optical events in seeing** are more familiar to us than those facts. We all have a more or less clear idea of how the picture is achieved on the delicately branched retina. The retina, however, is not equally sensitive in its whole extension, but just a very small part of it accommodates totally clear seeing, of shapes as well as of colors; therefore we move our eyes back and forth far more than we ourselves are aware of. We do not look at two things close to each other at the same time, but one after the other, however, in such a fast sequence that we are usually not aware of the change in the eyes' position; if, however, we want to see clearly **one** certain point, then we need to keep our look directed to it, which we would not need to do if perceptions were equally sharp in the whole area of view.

The circumstance, that we predominantly use only a small part of our retina and that we are caused to use always again the same one, is important inasmuch as the **retina possesses the property to become very quickly accustomed to a light signal, whatever type it may be, and then to perceive it as weaker, but to be doubly receptive for the opposite signal in the next moment.**

Similarly, like a dark night appears even darker to us after the bright flash of lightning, a dark area seems to be much darker to us than it actually is, if we look at it just after focusing on a point of bright light, and the other way around. In respect to colors, the opposite one always appears more purely to the eye after looking at **one** color, i.e., the one that complements the first one to white. So red is the more fiery the more it is surrounded by green, blue is related to orange, yellow with violet, etc.

This is the **place to discuss more in detail the way sensation of color is achieved with us:** According to the present state of science, it seems to be proved that the delicate ends of nerves which cover the back wall of the eye are of three different types; that means that some of them are sensitive to waves of light of longer length, some to waves of medium length, and some to waves of shorter length. It is presumed that the differences in perceptions of color are produced in that these different types of nerves are activated in different degrees at the same time. This hypothesis allows the explanation of many very conspicuous appearances, first of blindness to colors, then the fact that very different mixtures of light may effect one and the same impression of color, as for example white-appearing light can be produced to the eye out of all colors of the rainbow as well as by the mixture of two complementary colors.

If we take any red object, e.g., a red piece of paper, put it onto a white sheet of paper, look at it uninterruptedly for about half a minute and then quickly remove it, we will see at the place it was a greenish after-image, or in other words, according to the hypothesis: the nerves especially sensitive to red light waves are exhausted for a moment on some part of the retina, therefore they are further activated only slightly by the red rays included in white light; that is why the balance of the light effects, with which we perceived the paper as white, was disturbed. From this it follows that we see the paper colored instead of white for some moments with the exhausted part of the retina. The deducted term for this phenomenon is **successive contrast.** Besides this one, another important type of contrast is known—the **simultaneous** contrast.

If we repeat the experiment we just did and now pay attention not only to the created after-image but also to the surrounding area of white paper, we will notice that this, too, seems to have been changed: it appears darker and carries a slight amount of the complementary color of the after-image, that is the color we used in order to produce the after-image, in this case red. Maybe this is related similarly to the frequent experience that unequal sizes, put close beside each other, are easily estimated incorrectly, due to the ease with which we can perceive their difference; we are likely, e.g., to think of a tall and a short gentleman, standing close to each other, as taller or shorter than they actually are. Also it may happen that we estimate quickly-changing gradients totally wrong. If on a road a steep gradient abruptly changes into a very low one, then we will overestimate the difference of the gradients; we are going to think (if we have not trained the eye by long exercise) of the less steep area as level, maybe even rising, and then we will be very surprised one day, when we find water running along the edge in a direction opposite to the one we had expected. Maybe it can be explained in a similar way that even if we do not interrupt our look, meaning there is no following contrast, we think of light directly beside shade as brighter and of shade beside light as darker than both of them appear when mediated by transitions. In respect to colors, maybe not everybody will succeed in the experiment with the past-picture, because it is not easy to fix one's eyes completely to something, but on a winter's evening one can learn about the appearance of simultaneous contrast in the most comfortable way: while sitting close to the window and using the last daylight for reading, one may want to carry a lamp into the room; if its light hits one page of the opened book while the other page remains directed to the window, the first page will appear golden-red to us, the latter one will appear noticeably and intensively blue colored, nevertheless it still is illuminated by the same white light. As now in this way colors appear where there are none, so according to the same law **existing colors** experience a sometimes really striking change if put beside each other: **They appear more different from each other than they actually are.**

Now there are two qualities of our eyes left for discussion: as with the ear, lower sounds need to be louder than higher sounds in order to be audible, the eye needs to be hit with more intensity of light in colors of longer wave length than those at the opposite end of the spectrum, if they are to become noticeable. Therefore we see an object colored quite differently, depending on whether we look at it with bright light or dim light: in the first case we perceive mainly the reddish and yellowish shades of it, and in the latter case more the blue ones and violet ones. This differentiation is very familiar to painters. They subdivide the entirety of colors into two groups, which they call **warm and cool colors.** This name may derive from the fact that the warm colors appear more in the warm sunlight, the others more on the shaded side.

The division into warm and cool colors is nearly completely congruent with the separation into **forward-stepping and back-stepping** colors. The light rays' different abilities for refraction, which are demonstrated by the spectrum, are noticeable, too, in the media of our eye, and therefore we have to adjust its optical apparatus depending on whether we are dealing with light rays of a more or less separable type, under the condition that the sources are equally far away from us. If we put a blue thread beside a red one, we will not be able to follow with our eyes the way of the spun yarn of both of them at the same time, but we will need to adjust our eyes for the red one as if it were closer, and for the blue one as if it were further away. Then we easily get into the wrong belief that one of them really is closer than the other one.

Painters use a variety of optical illusions when they make us forget about the plainness of their board, and they know how to take many advantages out of these differences in perception of color.

The impression can be increased if the forward-stepping colors are kept brighter at the same time.

Up to now we have been fortunate enough mainly to follow scientifically proved facts in our remarks, but unfortunately physiology nearly lets us down completely with one of the most important questions still left, that is, **why so many of the possible combinations of color incur our displeasure.** Especially the pure colors are rather incompatible. So green does not look good beside yellow or blue, as Goethe has called the first combination nasty-cheerful, the latter one nasty-ugly. This judgement is hardly too strict. Imagine the uniform of a gendarme deprived of its gold braid, which separates the blue cuff from the green sleeve—it would look repulsive. (Only when they are very dark do green and blue get along fairly well, as the Scottish tartans prove.)

In general we should not put colors beside each other which are closely related to each other, but are not so similar that they might be regarded as merely shades of one and the same color, and between which neither a sufficiently distinctive separation nor a close uniting is possible. If it is unavoidable to bring them close to each other, one has to interpolate at least a narrow separation line with the help of another color going with both of them. **Which colors definitely go with each other?** First of all, one could predict the answer from the complementary colors, because it is known that colors increase their purity (or their saturation, as artists call it) by contrast; this conclusion is accurate inasmuch as they never look bad with each other; however, one is not likely to achieve with these the very best combinations. Only good taste, not science, is able to find these, and this is natural, too; for we remember that freedom is a major requirement for all higher levels of beauty, and therefore the wealth of beauty is not included in mathematically calculable forms and relations. As perpendicular crossings are less beautiful than some other angles, and circles are less beautiful than free curves, so the combination of complementary colors, which are closely related to each other in a supplementary way, cannot be regarded as the best one.

For the majority of cases, however, it is not yet sufficient that the chosen colors go with each other, but much depends on their being distributed reasonably, each one in an appropriate place. Therefore, we still need to learn about the principles, which guide decorative art in this respect. For our purpose we can learn more from it than from painting, for painting conjures up a pleasant **appearance,** while decorative art strives for pleasant **reality,** quite like in nature, and it tries to brighten up and to enrich things produced for human use with forms and colors that go beyond the absolutely necessary. This enrichment, however, has to stand in an intense and easily understandable connection with the object to be decorated, and it has to make it easier for us to understand the object in its form and essence. **Therefore the more expressive colors have to go to the more preferred places.** Under this point of view, one has to think about colors as being divided into three classes:

I. Gold, silver, black, white (we call these colors according to the common use of language) and instead of gold: yellow.

II. The saturated colors, that is the more or less pure spectral colors in their natural brightness.

III. The blended colors (those that are mixed with other colors), the dark and the pastel colors. The latter are those which are mixed with a rich part of white light.

An obvious example for a good combination according to these rules is the forestry uniforms: the fabric of the woods uniform carries a blended color, therefore it belongs to the third class. For decoration the collar and the braid have been chosen from the higher class of the pure colors—they are green. The undress uniform, whose fabric itself belongs to the second class, gets a decoration of the first class by its golden buttons.

The various types of army uniforms show more or less the same rule.

It is hardly possible to doubt the validity of this classification, but more interesting is **the question: which conditions provide certain colors with their generally acknowledged preferred position.** In part, **physiological properties of our eye** may be a decisive factor: the metallic gleam, white and yellow being the two brightest colors, may stimulate it, similarly the pure colors of the spectrum, but certainly there are other reasons, too. We will have to discuss these later.

The so-called **blended colors** are produced by decorative art in such a way that the pigments are not mixed but set beside each other in very fine lines or dots so closely spaced that the colored light coming from them melts into one solid color to the viewer's eye, if he is at a certain distance from it. Nuances of brightness often are created in the same way, in that white or black lines are brought onto the colored foundation, and it is obvious that in this way the harmonious impression of the combination is guaranteed.

One need not travel to the Alhambra to realize that, but may see it in many fabrics or woven tablecloths.

However, I need to point out that the rules of polychromy explained above are valid only under the presupposition that use of different colors is suitable at all. There is no doubt that a colored jacket may only be put together according to the polychromatic laws, but it is not said that a multi-colored one will always be more beautiful than a mono-colored one. Primitives in the tropical forest and aestheticians in the study room sometimes misjudge that, the former in tattooing themselves, the latter in wishing too much to recall the "colorful" middle ages.

Now we finally have reached the practical application.

If we start with a look at the details, at the single parts of plants, we will find that plants observe with marvelous accuracy the rules which humans also follow when adorning a thing with colors. **We immediately ask ourselves how this correspondence of the things nature shows us with the standards of our sense of beauty can be explained.**

I do not know any other answer to this than to refer to the proof offered in Chapter 2, Section 1, that our sense of beauty has been created according to the same laws of reason as all the rest of existence. It could be said as well the other way around: In nature the same laws rule as are decisive for our reason.

Hermann expresses himself in the same direction: "The general judgement of humans about the aesthetic value of a color at last follows everywhere in an unmistakable way the position of the color in the whole institution or the economy of the distribution of colors to the various fields or provinces of appearance in nature. **Nature, too, uses colors with a certain meaning or reason for illustration of the essence of the material things.** Generally every color in nature or in objectiveness rules a definite circle of things or appearances; whether in subjectivity or in human life, it is applied preferably to a definite circle of things, purposes and terms. This latter circle, however, always follows the first one as a sort of continuation or reflection. So, for example, the aesthetic value of blue color follows for us first the general importance or the character of the objective natural phenomenon of the sky. Here a wide field of empirical observation of the real existence and the aesthetic values of colors following it is opened up. However color actually is nothing but merely empty appearance to the eye, but we involuntarily put a reality into this appearance , and the appearance, too, is connected in a necessary and organic way to the whole nature of the things surrounding us."

Continuing our examinations under this point of view, far away from cold materialism as it is represented by Rossmaessler, we are going to think of our subject as even more interesting, for working according to a law stands higher than the chances of blind coincidence.

First of all, let us recall that the distribution of color on plants is not coincidental.

There are so many individual cases in which one can prove that certain colors have a definite purpose at their place, that one may say: **Every color serves a certain purpose. Where, however they serve more important performances of life, we need to encounter as well the more efficient colors, and in this way the most beautiful order is guaranteed.** The bark, which is solely covering and protecting, lacks bright and pure shades of colors; the leaves, which transform the raw matter into organic matter, are already one class higher. Flowers and fruits are supposed to attract insects and birds in order that they perform some essential services to the plant; therefore they are colored in a way that they especially appeal to their eyes. That is why they carry the same colors that we put into the higher classes, being such which involve our nerves for seeing especially well, namely the pure colors and the ones which are equipped with different sheens due to the special character of their surfaces. With seeds, even black is not lacking, which appeals strongly to the eye of the bird, which is supposed to distribute the seed, as well as to our senses, because it is so rare in nature.

Now it is no longer surprising that we see the first principle of polychromy, the distribution of colors according to various groups, according to the importance of the plant's part, observed so accurately by nature; in another respect, however, we notice **that nature acts much more freely than human art does. Often it puts closely related colors beside each other without fearing the impression of the vulgar,** and after all the yellow flower of *Ranunculus* on a green meadow, the blue forget-me-not at the green edge of a ditch, the green leaf of the water lily on the blue water surface have certainly never displeased anybody.

Let us take a look at **why things are possible in nature that are displeasing in art.** For an explanation I would like to recall an everyday observation: We recognize very easily and strikingly when the colors in somebody's clothes do not go together; hardly anyone, however, observes whether the colors of the sofa's cover fit to the suit of the person sitting on it at the moment. **So we are strict in judgement only in relation to combinations of colors if the things which the colors are fixed to suggest combining them into a unit. The freedom of nature, however, is beyond such uniting.** We learn immediately that this is indeed a great part of its magic with which it avoids ugly things, when we try to use plants for human works. When we unite plants to a bouquet or in a bed, we are not allowed to combine rose and red lily, *Pelargonium* and *Phlox* without punishment. The most dangerous ones are the carpet beds with their sharply edged colors moved close to each other. Nature takes care not to take risks like these. It does not give sharp borderlines, but leaves us the total freedom to combine out of its abundance what we would like. **Then it is noticed by the observant eye when something happens to go together extraordinary well, and we unite those things; however, we overlook easily and often the less beautiful.** I take the forget-me-not into my hand, and I enjoy the wonderful blue, the nice yellow in the flower's center and the delicate white edge, which separates accurately the two colors; I also take into account the beautiful shading of the colors as it takes place between the youngest pink-red and the older blue flowers; but one does not regard it as unpleasant when the blue flowers really do not go with the leaves of some lush neighbor plant (the dull green of **their own** leaves provides a very good background to their much brighter and shining blooms).

There is a second example that may be better known to the reader: the blue-green of the *Vaccinium* stems, stripped of leaves in late fall, and the yellow-green of the mosses in the pine forest definitely do not go together, but who is going to notice that? Maybe someone who has read an essay like this one or intends to write one himself. Over here the two plants mingle with each other hardly noticed, over there they are separated by ferns and dying grasses, and in the distance the small tufts and carpets melt into one even green blanket to the eye, so that we enjoy our dear pine forests very much with this adornment, without criticizing in the least the details which are noticeable in the foreground.

By the way, the various shades of green are colors especially incompatible among themselves. Painters know that very well, and that is why we will certainly find in every exposition about ten paintings with brown trees before we will encounter one with green trees; if this one is not made by a master, the audience usually will not really look at it at all, it will soon shout: "pure spinach," and move on to all the other colors, but not to green.

If, however, nature always succeeds in satisfying us, although its main color is that obdurate, it must have even more advantages than the ones discussed above relative to the artist, and this is true indeed. It defeats us by its **way of mixing, combining and separating.** In what a marvelous way it uses everywhere the trick, always difficult for the artist, of producing mixed colors by putting other colors in delicate lines or dots on the rendered ground.

If it wants to shade the green into white, into black, into blue, red or yellow, then it will cover the leaves and stems with delicate hairs or glands, or it will put a network of veins in the necessary color on it; and in the whole: how rich but cautious does it embroider the carpet of the meadows! Never will it let one color appear all of a sudden; first it brings a new color in few dots, then amply, here completely dominant, there vanishing and giving space to another one. It knows to manage so that all the thousands of color shades that the meadow runs through from spring to fall will never show even a single disharmony. The steel-blue of the Fuchsschwanz [*Alopecurus agrostis*], the yellow of the *Ranunculus*, the delicate pink of the *Cardamine pratensis* and the light violet-green of the

Honiggras [*Holcus mollis*], the brownish blue of the *Molinia carrulea,* the red of the carnations [*Silene spp.*], all of them receive their spaces and every color finds its transition to the other one arranged.

Quite similarly, the marvelous play of colors is created on a water surface moved slightly by the wind, the special color of the waves and their different reflections weaving themselves into the unified main color of the surface.

Being a master **of the most delicate shades and the most sensitive combinations,** nature always takes hold of this art, where it wants to show the most perfect beauty, where it **intends to have an effect by shape and by color at the same time;** for unblended colors purely put beside each other attract too much attention to themselves and may often damage the impression of the whole. So the forest snipe is without doubt more beautiful than the parrot, and nature knew very well what it did when it denied flower adornment to the beech and the oak, which it had given already to *Salix caprea,* to *Prunus serotina,* to *Daphne mezereum.*

By the way, the delicate transitions are only partly tied to the thing itself, but often are created only by the act, that **leaves, water and air do not leave the light uncolored which has gone through them or is reflected by them. They give to it a mediating nuance to which the landscape mainly owes its unified character.** So for example the cold white-gray of the trunks of the beeches does not go well with the warm green of the beech's leaves, but one will never be able to notice that in nature, because the bark receives a warmer, more fitting shade out of the top of the tree by passing and reflected light. A picture I possess proves that this is really a very important fact. This picture is the accurate portrait of an old beech, marked by a marvelously gnarled and striking light trunk. The artist has paid special attention to this trunk, and maybe that is just why it happened that he painted it in the color which is typical of the bark (white light provided), but not like he saw it outdoors. The consequence now is that the whitish trunk is standing somehow strangely in the landscape. This is really the reason for the failure; I learned about it when one day I happened to wear blue glasses and saw through them the trunk not garishly separate any more, and therefore saw the whole landscape picture colored in the most harmonious way.

One can ask, however, how could the painter make this mistake, to paint something in a different way from how he saw it? I think it can be explained very easily: When he made the sketch in the green forest, his board gained the same advantage as nature, in the way colored light poured itself onto it, and this gave his colors a harmony it had to lose again in white light. The same experiences can be made with water. Here as well the beautiful transitions created by reflections are the ones that exclude the possibility of a disadvantageous combination between the blue, typical of the water surface, and the plants' green. The mirror image of the latter moves mediating between both of the colors. One can hardly compare often enough theoretical thoughts with the appearances of reality. In respect to the last sentence I found the most beautiful opportunity to prove its correctness just after I had written it. After a heavy rain I saw a number of small water puddles on a medium sized pasture, each of which had a totally different color according to the part of the sky it reflected. The reddishly shining ones were the prettiest ones, no doubt, but also the bluish ones looked very good on the green pasture, although there was not even the least transition color by reflection visible, to which I had just attributed the harmonious effect (probably justified for the majority of cases), for the grass had been cropped by the cattle too much to have any of it reflected. Then I soon realized: **the blue water and the green pasture went together well because their intensities of light were so very different.** One might remember that I believed the reason for the bad impression of certain color combinations is that we are embarrassed by them, because we do not know whether we should perceive them as different colors or as shades of one and the same color. In this case, however, the great difference in light intensity makes the distinction easier for us and removes the disturbing haziness.

It became more probable to me that this explanation might be the right one when I imagined the pasture as being part of a landscape painting. Then I had to tell myself that certainly the colors on the canvas would not have been allowed to be represented in a different way without becoming ugly. In a painting exposition in Breslau I could study the correctness of this assumption in a painting which was very beautiful in every other respect. The painting represented a lake in the Alps with a marvelous green water surface. At one place in the background, the water surface obviously had been moved by a breeze coming in from a tributary valley, and there the waves reflected the clear blue of the sky. Only a little imagination is necessary to visualize how beautiful this must have been in reality; in the painting, however, the two colors related very badly to each other, obviously because the painter had not had the same differences in light intensity at his disposal as in the actual scene. In a similar way, it will be possible to explain that the tree tops fit so nicely to the sky's blue; while it must be noticed that usually the treetops are seen against the horizon, where the sky mostly shows colors other than pure blue, while on the other hand, the leaves of nearby branches get a yellowish shade in the light filtered by foliage.

Whenever it is necessary to unite contradictory things, nature has another important means at its disposal: **the atmospheric perspective.** The air is never totally pure; it always contains, even in very small amounts, water vapor and other small particles which have a more or less noticeable influence according to their number and size and to the actual position of the sun. Put in front of distant things, the impurities reflect a part of the light rays back to the distant things, and let other rays which they receive from other light sources, the blue sky or the clouds come into our eyes instead of those rays. It is due to this relation that we see the distance shaded in such a wonderful unity, after the same law of decreasing clarity of the colors; not only the distant background, but even much closer distances than we usually think. It is only necessary to look at a painting from Boecklin, who disregards deliberately the aerial perspective in respect to the red color of the garments, to realize the difference between the painting and the color combinations possible in reality and to become aware at the same time about the advantages of reality.

The means we learned up to now are sufficient for nature to represent the rich abundance of harmony and beauty we enjoy everyday; sometimes, however, it takes as its task **to affect us more deeply by more sharply tightened contrasts.** This is especially true for the daylights of morning and evening, which start and finish the day in solemnity. **However, how is it possible that we never miss the harmony, even in these pictures that show the strongest contrasts?**

On the plain (I do not know enough about mountains to be able to discuss their situation in detail), the most beautiful light effects are probably found in the pine forest at dusk; let us take these for example in order to prove in them the principle of the visual effects. Whoever sees for the first time how the tops of older pines dress themselves with marvelously glowing colors in the light of the setting sun will probably not long for the Mediterranean pines of the south, being full of deep respect for these honorable trees just in front of him. How is it possible that the usually modest garment of our friend, the pine, can all of a sudden appear so richly decorated?

It is very easy now to deduce the explanation from the physiological contemplations above, and the details that might have seemed to be too much in the beginning will now be favorable. It is sufficient now to remind the reader that rich light comes in useful for warm colors while dim light comes in useful for cold colors. We then realize how successive and simultaneous contrast deeply increase this difference in coloring, and finally we notice the reason in the forward-stepping and back-stepping behavior of the contrasting color groups, which makes the shaded inside of the forest step further backward, and in this way gives greater spatial differences to the landscape image than it really has.

The simple example gives us an answer to the question above: the color contrasts of the evening light belong only in a small part to reality, but they find their creation considerably by qualities of our sense of sight, and therefore it is only natural that they are pleasant to our inner being, that they please us.

Also the **winter landscape** provides opportunities for instructive observation in this direction. However paradoxical this may sound, especially the winter landscape gets its main attraction from the **color contrasts** between the yellowishly or reddishly appearing snow parts and the shadowed parts, which have colors going to blue or violet. Here we find the simultaneous contrast in its most beautiful formation.

One often is unjust towards the colors of the winter forest. The song praises the honest fir for its winter green, but who thinks of the beautiful red-brown which old alders get from the new blossoms and the seeds from last year, the marvelous brown-violet which veils the tops of stands of beeches, if one looks at them from far away. (The coloring stems from the tips of the scales of the spring buds.)

The enthusiast for the spruce as the tree ideal in every respect might want to look at the winter landscape quite impartially; I think a nice day in December, then, is going to heavily shake the rooted prejudices. Quite the contrary, the low forest shapes the winter landscape very friendly, because it has the most young trees whose bark is light and vividly colored. At the same time, it is not necessary to do without the green of conifers. One can include these in larger or smaller groups. If they are applied sparingly, they will not appear as gloomy as in large amounts, and their color gains in vividness by contrast; by their proximity, the warmer shade of the hardwoods will increase.

In one respect, however, the conifers are incomparable. In the middle ground of the landscape, their strong colors are not yet changed very much by the aerial perspective, and therefore they cause the background to appear lighter and more delicate by the contrast.

But I find myself anticipating the applied second part of *Forest Aesthetics*. Before we reach that part, there will be some more contemplation of single fields of the beauty in nature.

Chapter Three.

Stones as Adornments of the Forest

A forester is not able to regularly visit art exhibitions, he is not often seen in the theater and in concert halls, and seldom has the time for enjoyment of the works of poetry by reading good books. But the forest is to him art exhibition, concert hall, theatre and library, all at the same time. Besides the forester, there are many other people in the same situation. Recently I found the idea that **enjoyment of nature may compensate the modern human for enjoyment of art** expressed in very nice words: "To our educated eyes and our scientific understanding, the landscape has not remained a colorful image, but it presents itself to us as a **self-grown work of art,** in which the relations between cause and effect are expressed."

From the poetic point of view, every year dramatic events develop. Every day from morning until morning again, every change of weather and the seasons, renews in front of our eyes a regular sequence of sometimes lovely, sometimes sublime spectacles; but everything we experience now, how small it appears in comparison to the immense shakings to which the earth owes its wrinkled face of today. We foresters are appointed to work there where the traces of those struggles are especially clearly visible—that is why we are interested in the wrinkles and the folds and the old scars.

A forester who is satisfied just to know what will grow on the soil he cultivates—for example that beeches grow extraordinarily well on shell limestone soil, that on windblown dunes of quartz sand only miserable pines will grow, and that on the mud of the floodplain ashes and oaks will shoot up—is comparable to a pedant who looks at the Rheinfall [falls of the Rhine] only as a source for hydro-electric power.

As geology is the foundation of the teaching of habitat and site, it should be as well the foundation for forest aesthetics. Being aware of that I would have loved to dedicate a contemplation to the nation's soils; to the lovely waves of the ground moraines, where I

I.

Geschiebe führender Wildbach

Gemälde von Karl Hasch.

live, to the Zobten Mountains, this picturesque symbol of middle Silesia, which has been built up by granite, gabbro and serpentine, to those round hilltops whose seemingly infinite sequence guides the view from Wartburg Castle into the distance. But to describe so marvelous things is still "beyond our ability."

In order to train my eyes for the contemplation of the earth's surface as a work of art, I started with small things, studying the **layering of the deposits and of the moving stones;** I chose those because they are partly still in movement today, and the rules of their layering are relatively easily recognizable, and because the recognized rules can be applied in the practical part.

Because there are only a few forest districts that lack stones completely, I hope that the following examinations are going to be interesting to at least some readers, if not to all.

Inspired by Moltke, who praises the artificially constructed trout creek rich with rocks near Peterhof, one day I tried to give more change to a small bordering creek by rolling some field stones into it.

The small rapids looked pretty nice, but soon there was a thunderstorm, whose floods whirled everything around. Only now, when comparing my work with the one of nature's forces, I realized how miserable my attempts had been. This experience first made me look for advice in the **literature,** but in the books accessible to me I have found only a few things about this subject.

References on stone groups created by the art of gardening usually showed only how not to do it. However, it is also hard to learn from nature. When one goes into the mountains in order to study the appearances, first one is overpowered by the infinite variety of pictures. But whoever is seriously looking for "the resting pole in the flight of the appearances" (Schiller) will notice a certain principle, which will become more evident the larger the relations are one is looking at. For example, v. Domaszewski writes about the area of the Danube where the detritus will remain mostly in the bed of the torrents and will rarely come down into the area of the mountain rivers. "The coarser, fist-size stones will be carried by the stream at most down to Pressburg, gravel reaches Ofen-Pest, then simple sand down to Widdin, finally silt and clay down to the river's delta in the Black Sea."

That this must be true is obvious even to the lay person, but much more difficult is the explanation of the shaping in the **upper part of the river.** If it has changing gradients, then one should think that all the stones, large and small, will be found all together where the gradient is the smallest—because also the large stones should stay there most often, where they can resist the current the easiest.

But actually the opposite is the case, for one can find the most and the largest stones on the steepest places. I learned to understand this mysterious phenomenon by a work of F. Wang, who explains:

"Every increased transport of material, other conditions being equal, decreases the average speed of the water, therefore also its power of pushing and the erosion."

Figure 2 may explain how I visualize this action.

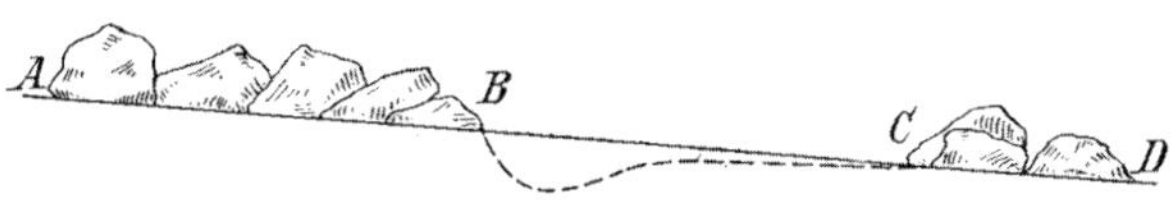

Fig. 2

The dotted line represents the new creekbed formed by "erosion."

On the even running creekbed section AD coarser stones may have been deposited between A and B. Then in this section the bed will be protected. Below B the successful attacks of the water continue; first the bed directly below B will be damaged heavily, so that a water-hole will come forward; further down it is less severe, but still sufficient that the bed becomes nearly level, until still further down there are again some stones, which are a restriction for deeper washing-out. I have no doubt that this explanation is true now, after I found it. Its correctness can be proved by every walk along a ditch with alders. There one will find the strongest gradients always between trunks and tree roots, while the assumption is excluded that the alders would have been able to settle down only where the water runs more quickly. Here it is clear that the shape of the creek was determined by the restrictions of flow, not the other way around.

Also in another way the rules of "sorting of material," which are valid for whole areas of rivers, are violated partly in rocky torrents, for it is often that one finds the heaviest blocks put in front of the small detritus. This contradiction I would like to explain by saying that the larger detritus can be moved down the valley only during a **mudslide.** This, however, follows its own special rules. Wang describes this very graphically: ". . . in the case of a mudslide, when the water has taken on a mushy consistency with a higher specific weight due to the suspended soil, extraordinary large and heavy stone blocks, often even rock parts the size of a house, are carried down the valley easily. Like a closed phalanx the mudslide slowly moves forward, in which the mass of the detritus often is more than that of the water, and in which the stones first are lying close to each other in wild disarray and are carried away touching each other, mixed with mud; the mudslide sweeps away with it everything that stands in its way down the valley. Within this mass, which is extraordinarily able to carry weight because it is specifically heavy, a sorting of material takes place soon, such that the largest detritus tends to hurry forward due to the increased power. The detritus masses of the mudslide slowly get free of their intense contact, so to say, and grade themselves in a way opposite to the individual transport. Now the coarsest material is hurrying first, nearest to the sole, and the finest material is staying behind and can be found in the current on top."

It is hardly necessary to mention that transition forms do occur.

As soon as the mudslide has come to an end, the individual transport again takes place and rearranges the stones in its fashion. However, this does not happen as quickly as it is written. The small light stones that stayed behind would be expected to move first. **Very often, however, they are forced to stop** on their arduous journey. The causes for this stay are many and various. The most common cause is that the smaller stone resists the current by being protected by a larger one or is hurled up to the shallow edge. Also, its own direction against the stream can delay its travel, if it has gained a position in which it offers the least target to the stream. This, for example, takes place with cylindrical stones when their longitudinal axis is parallel to the stream's direction. Smaller stones also find temporary rest directly below large stones and as well at wide places of the river bed, because in both cases the current is reduced.

More often the wandering stones hinder each other. This may happen in three different ways. If a large stone has settled down in the watercourse, then others can be delayed by it. Furthermore one can observe that flat stones move up partially over other ones, as shown schematically in Figure 2, Page 57, at B. Flat stones like these then are very stable and hinder further masses in their movement. More rarely, the stones jam themselves like a vault, but if so, then always in a way that the closed side of the arch is directed up the hill. In this way waterfalls and rapids are created which are usually especially beautiful, because the water veins falling down the stones are directed towards one focus, they cross themselves and unite again, then separate again.

It is wonderful that garden artists usually do it the other way around. With the help of a lot of cement the arch is directed downslope to the valley, contradictory to all rules of construction, as if one had preferred to take the mouth of a small jug for cream as a pattern instead of nature.

Round stones follow other rules. They are found in two different positions—often crushed or caught between larger sharper stones, more often, however, at the edge or in shallow areas, between much smaller flat and sharp stones; for large balls hardly ever stop again after they have begun moving. Their own power of movement snatches them away from the influence of the current.

Unfortunately I have not yet had the opportunity to test the correctness of the rules developed above with glaciers and **at the**

seashore, for I have not yet seen glaciers, and to the seashore I have been only in earlier years when I had not yet started to pay attention to the settling of stones.

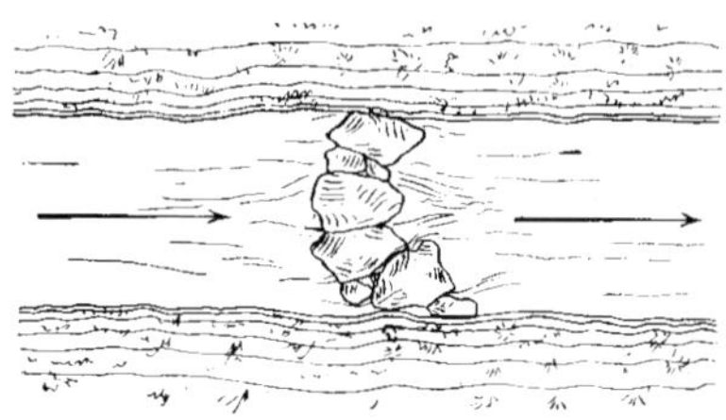

Fig. 3

Nowadays, however, I very often have the opportunity to study the excellent painting of the Baltic Sea beach by Eugen Bracht in the reading room of the Reichstag [Parliament building], and see for myself that all the elongated stones (about 30 of them are represented) without one **single** exception are directed with their longitudinal axis from the shore towards the sea. This is quite understandable, for as long as a stone keeps a direction parallel with the breaking of the waves, it will remain a toy of the breakers and will be moved forward and backward; this will stop at the moment it resists the water's attack with its narrow side directed towards it.

Having discussed in detail stones in and near the water, I can stay briefly with **rubble stones,** which have not yet started another wandering.

Stone rubbles not moved by water also move down the hill to the valley, but the travel is slow, and therefore they have time to make themselves comfortable. That is why they mostly lie on their **widest surface** and are always **bedded into the ground** more or less deeply, if there is soil beneath them. If, however, other stones or even solid rocks form the ground, or if trunks hinder the wandering, then the stone will always move more slowly than the soil, which moves down the valley, too. Holding up the soil then, **the stone will form a step.**

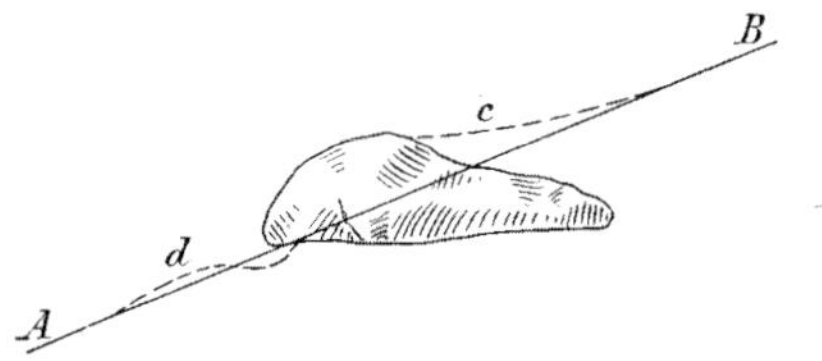

Fig. 4

AB is the original gradient of the soil. The dotted line at c suggests the new surface line caused by a slide, while the new formation at d is caused by wildlife.

One can find very varied pictures at fields of rubble whose surface in its entirety is covered more or less densely with stones. As humans move heavy loads on underlain cylinders, nature helps itself, too. Large slabs of stone prefer to start wandering **on a movable base and with the least possible surface for friction** (Figs. 5 and 6).

Fig. 5

Fig. 6

The wandering stops whenever they meet solid obstructions. Then they will hinder each other and quite often a formation **with the character of a flight of stairs** will be created (Fig. 7).

Fig. 7

Stones which have traveled mostly by water but by now have been on dry land for quite a while, e.g. the **findlinge** [erratic boulders], follow the same rules in respect to sinking into the ground and slowly migrating down the valley.

However, there are some other stones that never have traveled. Separated from the bedrock by irregularly progressive weathering, they happened to be in **equilibrium** or nearly so, so that they were not able to overcome the frictional resistance of the base and had to stay in a seemingly uncomfortable position. Such formations are the ones which natives usually show to the strange visitor with a particular pride.

Every reader will be aware of the fact that foxes, rabbits and other wildlife like to dig under and between stones, just as plants like to colonize near stones.

Lichens and mosses increase the most the impression that a block has been in place already for a long time.

In his poem "Atmosphere," Goethe accuses:

"I have to see all this with my eyes,
but it cannot be really understood."

This is true for the settling of stones as it is for the layers of clouds. Just as fitting is the poet's comforting word:

"In order to find yourself in the infinite,
you need to differentiate and then unite."

I hope for the reader the uniting of the single appearances in nature into a unit ruled by laws will be easier than the differentiating has been for me!

Whoever dares to train himself to separate and unite, not only mentally in admiring nature, but who follows its models and tries **to use raw stones for embellishment purposes,** he will be able to recognize the difficulty in its full amount.

Chapter Seven (Part II B) will try to make the difficult task easier by some practical hints. This chapter was meant only to increase the pleasure in beauty as nature provides it, and especially the understanding of its stone language (*saxa loquuntur!*) [the stones talk!].

So one remark of Selenka, whom I already quoted at the beginning of this chapter, may come true:

> ***"Adornment is nothing else but a natural language, intelligible to the generality, which is suitable to tell the neighbor metaphorically about our merits."***

This is not only true for human adornment; the forest also tells a lot by its adornment (stones are part of it) to its neighbor, the forester. May I have succeeded in encouraging a little bit of the understanding of this "natural language."

The picture at the beginning of this chapter reveals that the artist has taken nature as a precise model and therefore it is very educating. We see the waves of the Giessbach [Giess Creek] foaming over stones which protect its bed against erosion, while in the foreground it reaches a more level part free of stones. On the right bank we see the T-shaped block held back by lying on top of smaller stones. Gravel has maintained itself, being protected against the strong current by the block. Coming from both banks rubbles of rocks push themselves downstream, all of them laid with their longitudinal axis in the direction of the steepest gradient. Those that have not yet been uncovered by the Giessbach are sunk into the soil and form ledges like stairs.

Chapter Four.

The Aesthetic Value of the Tree Species.
1. Preliminary Remarks

The plant world is, so to speak, the garment of the earth. Being beautiful in all parts, plants come together to marvelous unifications, each one of which has quite a special character. It is the task of forest aesthetics to appreciate the forest plants aesthetically and individually and then to examine why they fit together so well into the harmonious entity which we admire and love as the German forest. The following pages are intended to provide a basis for that, which is limited to woody plants.

Following the forestry custom, I will begin with the hard hardwoods, followed by the conifers and the softwoods, finishing with the shrubs. A special paragraph is dedicated to some foreign species important to forestry. If I dedicate more space to the varieties, specialties and forms of forest trees than some of the readers feel is justified, then one might want to believe me when I say that I am already practicing strong self-discipline in leaving out much of what I am personally interested in.

Those members of the Silesian Forest Organization who have seen the 90-year-old stands of *Quercus robur,* grown out of Russian seeds, whose straight, vertical trunks distinguish them from the oaks of minor value in the neighboring stands in the city forest of Gruenberg, will know how to appreciate the aesthetic value of good varieties and sub-varieties.

Less beautiful varieties are important, too, because they "reveal the idea of the thing." I owe this expression to Oersted who writes:

"Nature carries out each of its ideas in innumerable varieties and works, whose bringing forth takes place in incomprehensible times; in the entirety of them all the whole idea is supposed to express itself. Just as a philosopher forms for himself one main idea into the most different shapes, just as a performer does the same when he varies a text, so then does nature, but in an incredibly greater diversity. Every unified being (individual) is such a characteristic implementation of the main idea of the object. But the rich nature does not limit itself to showing us implementations in which the ideas stand in front of us practically finished; no, it shows to us with countless changes of the relations of finiteness, changes which a partial observer will call the imperfection most obvious to the

II.
Postel. Susannen-Eiche.

eye, changes which, however, the one who has made the perception of nature his own in the high amount which should be developed among all human beings, will find to be destined to reveal the idea of the object in all its richness to a mighty clearly-seeing mind."

2. The Hard Hardwoods

Oak.
[*Quercus petraea* (Sessile Oak)]
[*Q. robur (Q. pedunculata)* (English Oak)]

The first place undoubtedly belongs to the oak, because in its old age it attains **the character of the sublime.** It owes this character to its proportions and the strong structure of all of its parts, from which you can read that every single trunk follows, in addition to the rules of the species, some intrinsic rules of its own.

Merely in the position of the buds it reveals its character with a mind of its own; the buds are all arranged in a spiral, but are often very different from one another in size. This is also true of the growth of the stems, the branches, even the roots, if they are visible above the ground.

"*Tum fortes late ramos et brachia tendens huc illuc,*" the poet praises them. Here and there it stretches its mighty branches as it likes, but it keeps its equilibrium—*media ipsa ingentem sustinet umbram.*

Esculus inprimis, quae quantum vertice ad auras
Aethereas, tantum radice in Tartara tendit.
Ergo non hiemes illam, non flabra neque imbres
Convellunt: immota manet multosque per annos
Multa virum volvens durando secula vincit.
Tum fortes late ramos et brachia tendens
Huc illuc, media ipsa ingentem sustinet umbram.

Virgil. Georg. II. 290

[First the oak, which stretches out a crown to the upper
World, as large as the roots to the infernal regions.
Therefore no storms, no blasts of wind, no rains tear it:
Unmoved it remains and by its toughness outlasts many
Generations, many ages of man; then stretching its strong
Branches and arms here and there, all around from its center
It supports great shade.]

The **impression** the senses perceive immediately by its presence **is increased by that which we know about it.** We know about **its power of resistance against the elements.** While it declines to whisper in the soft breeze, it raises its voice to a roar in the storm and fights to resist it. We know to interpret poetically to its favor the fact that the immense power of lightning tests it so often. We find it coming out of the struggle not without honorable scars, but ready to unfold its **vitality** even more immensely in thriving new branches and new tops, this same vitality which the poet uses as a metaphor in order to compare it with the flourishing time of his nation. **Of the nation,** not of a single man, for how much does the single mortal diminish if he compares his **lifespan** with that of the oak, if he counts back in the rings of the fallen trunk, whose diameter rises above his chest height, to see who may have been emperor in the German countries at the time when the wild boars hid the acorn out of the richness of the harvest into the virgin soil of the old selection forest; if he asks himself apprehensively where we shall have arrived, where the children and grandchildren will be, when out of one of the acorns which we now put arduously into dugout strips, a trunk will have grown equal to the one now lying in front of us.

Some observations come additionally, of a minor kind, but suitable, however, to intensify the impression: often we see the oak tower above other tree species. It seems to protect them royally. It is the last one to green up: it follows the others as the most distinguished one. It uses minor plants as its garment, in order to moderate the all too great roughness of its bark with magnificent color and gentle wrapping. It offers food and shelter to the numerous wildlife of the forest, and finally it is ennobled by the use humans determine for it, for its wood is not for sale for common purposes.

Its foliage—from the decoration of the lucky hunter up to the honor garland of the homecoming warrior and the honorary portals for princes and kings—represents the laurel to us Germans.

Painters and poets of famous as well as infamous names have always concerned themselves with showing the advantages of our tree in a favorable light and with carrying understanding for it into wide sections of the population. Not everybody has a naturally open eye, ear and heart for the beauties that surround us. Only the artist can open up the meaning for most people. In favor of our tree they have not lacked good intention. The poets even have approved—Klopstock first—to claim the widespread tree species as being exclusively affiliated, inherited and in ownership to us Germans, and none of its advantages has been overlooked by them. The forestry journals of the first third of our century are particularly rich with partly excellent poems of this topic.

The painters are even more hard working with this matter. Their brushes are not able to cope with every oak, the least with the ones the forester prefers to grow: the slim trunks of the closed high forest and the rounded tops of the well-groomed canopy trees in the composite forest; but they make better use of those shapes which used to grow in the old selection forest and preferably in the grazing forest. Already Gilpin realized that when he praised the picturesque beauties of his home country's oaks, a wide forest area on the banks of the Avon, where one hundred years ago wild horses still roamed.

He recognized those oaks, which had grown out of the reach of the freely grazing ponies' teeth, as "... the most picturesque trees which one can see. They seldom get tall trunks like other oaks in richer soils, but their branches (crooked wood for the ship-builder) usually twine themselves around each other in the most picturesque forms. Anyway I think that the poorer the soil is, the more picturesque the tree will be—that means the more beautiful is the branching it will form."

"Moreover the oaks in New Forest are not overloaded with foliage like some others are in richer soil. An overload of foliage spoils all form. As well as in the opposite case, if the foliage is too thin, the tree will look spoiled, miserable and shrunken. A tree is perfectly picturesque if it has enough leaves to form a mass, but not so many of them that they are going to hide its branches. One of the greatest beauties of a tree is its branching. This has to shine out from behind the leaves every now and then, even if the tree is full of leaves."

I am convinced the dear reader will not discover anything special in the quotation above, but think of the expressed thoughts as being rather self-evident. However, it is not like this at all. We think of a theoretical contemplation as very plausible if it is presented to us in clear, intelligible words, but put in front of reality, the untrained eye does not always succeed in recognizing without doubt the picturesque beauty and enjoying it visually.

Even dying oaks can be picturesquely beautiful. Often it has been tried to derive architectural formations from forms of plants. Whoever looks at the withered branches of the oak with some imagination will probably be able to recognize some monsters—snakes, dragons, dogheads, as if they were gargoyles, or the strange decorations of Gothic buildings.

Next to the structure of the branches and the distribution of the leaves into great masses determined by it, it is the **arrangement of the leaves** in detail on the branch, and further, their size, shape and color, which are decisive for the appearance of the tree.

The position of the leaves of both our native oak species shows a far-reaching and aesthetically important difference.

The difference in length of the petioles is generally known, but it seems not to be considered enough that the surfaces of the leaves, in respect to position and shape, are influenced essentially by the length of the stem. For the leaves of *Quercus robur,* attached shortly to the branch as they are, are not at all or only partly successful in getting into the most favorable position towards the light, while *Quercus petraea* needs only to let its long leaf-stems turn around in order to have its foliage completely spread towards the sun. One can therefore say in a certain meaning, like G. L. Hartig did, that its leaves are standing alternately along the branches in the style of beech's leaves.

While the first type of arrangement shows to advantage mostly on older trees, in adorning every short-shoot [spur] of them with a well-arranged, delicately shaded bunch, the leaves of *Quercus robur,* spread out in a plane, look especially good on younger trunks which grow mainly long-shoots.

As with the position, so it is with the **shining of the leaves.** Its absence as well as its presence may be understood as an advantage, depending on the circumstances. The shining of an object is caused by the reflection of part of the arriving light beams as uncolored white light. Vivid color effects can therefore be achieved only with non-shining bodies or with those that reflect the light as shining light only in a direction not occupied by the viewer.

The duller leaf-rosettes of *Quercus robur* therefore can take every direction of illumination, just like a watercolor painting, while shiny foliage in the sunlight, seen from some distance, may look as mis-colored as the oil-paintings of beginners, to whom the exhibition commission has provided a place with bad, unfavorable light in order to grant to a Mackart or an Achenbach those lightings which reflect the shine of the plain board without damage into any angle of the room, while it makes the spots covered in the pastose-way look like a precious stone, a full moon or a wave coming out of the general gloom (the common layperson says "blobs of paint," when he steps closer in order to see how it is done). So often a branch of *Quercus robur* out of the half-shadow of sparse pines shines brightly far through the stand, and in this respect is quite similar to the beech.

By the way, there are numerous transitions between the leaf shapes of the two oaks, and *Quercus robur* with very shiny leaves is not rare at all. Here I put in the pictures of some leaves, collected by me in Postel and vicinity (Fig. 8–16).

a
Fig. 8

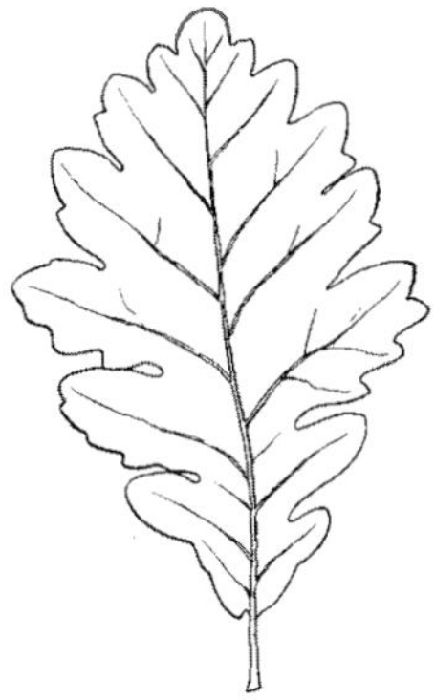

b
Fig. 9

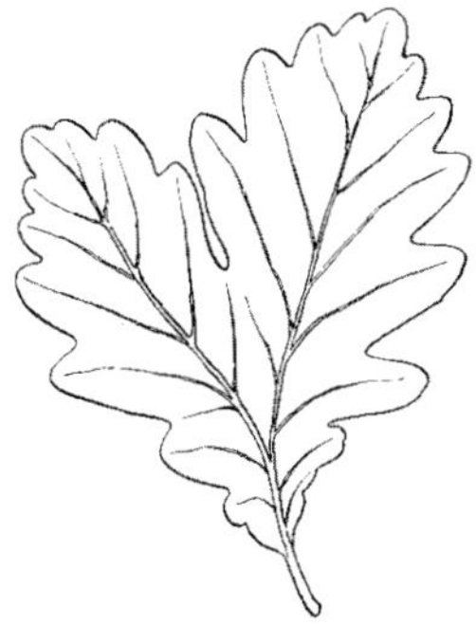

c
Fig. 10

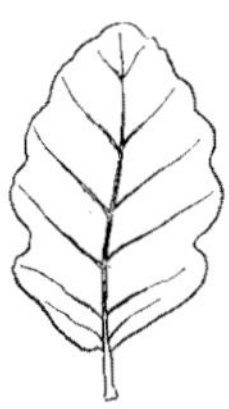

d
Fig. 11

e
Fig. 12

f
Fig. 13

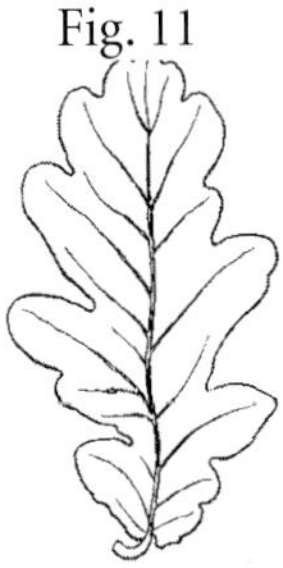

g
Fig. 14

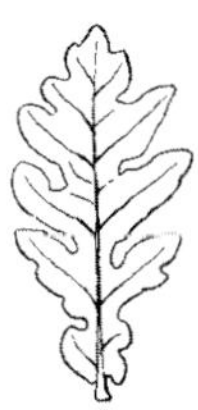

h
Fig. 15

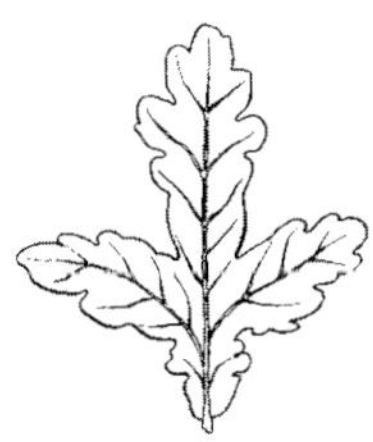

i
Fig. 16

The first leaf of a *Quercus petraea* (a) is odd, because it shows a transition to American leaf shapes in its sharply ending tips. The next one (b) is delicately doubly lobed. The leaf at (c) has a forked center rib, and those at (d) and (e) are similar in shape to alder leaves and willow leaves. Leaf (f) comes from a hybrid form of *Quercus petraea* which is marked like *Quercus petraea* by the leaf stem as well as by the multiple fruit, while the leaf itself has the definite character of *Quercus robur.* The most common form of the leaves of *Quercus robur* is shown in (g), while (h) can be found more rarely. G. L. Hartig felt obliged to distinguish these delicately deeply lobed oaks as a special variety (*Quercus altera tenerius dissecta,* Raseneiche) [Lawn Oak]. I took the leaf in form of a cross (i) from a Johannistrieb [June shoot—Johannis is in June]; it is a strange play of nature. The last leaf, shown in Fig.17, is special by failure of the upper tip, due to the stunting of the center rib, but even more special, however, because of its history.

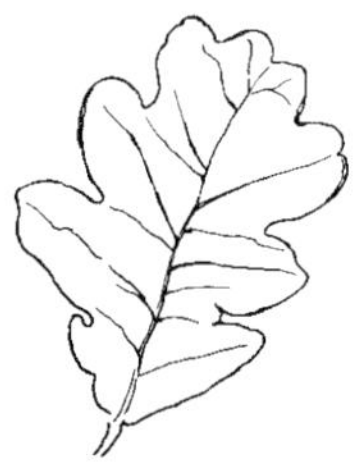

Fig.17

It was in fall, 1863, when Emperor Wilhelm, at that time still King of Prussia, held a court hunt in the Lesslinger Heide [heath]. Shortly before the beginning of an announced drive on wild hogs, His Majesty called for Mr. v. Meyerinck, who was next to him, and asked him whether he had ever found an oak leaf with two rounded tips in the forest. Mr. v. Meyerinck answered in the negative, saying that he had not yet paid attention to that. "Imagine," the Emperor then said, "I just found such an oak leaf; it is the second one I have seen in my whole life. However, I am going to tell you why I am interested in that: One day when I was with my siblings at my father's place in Sanssouci, he said to us children; 'Go into the park and see whether you can find an oak leaf which ends in two rounded tips. That would be interesting to me, and whoever brings me such a leaf will get a reward.' We siblings hurried away and looked very eagerly below the oaks. I was so lucky to discover the first leaf with two oval tips, and cheering, I ran up to my papa. I met him on the terrace and gave it to him. Very good, said the king, now I am satisfied; now I am going to establish the Red Eagle Medal with Oakleaf. You, Wilhelm, are going to receive your reward. —But I never got this reward," said His Majesty, and laughed.

The king gave the leaf "with two rounded tips" to Mr. v. Meyerinck, asked him for paper and pencil, and drew a common and a two-tips leaf.

It is due to the kindness of Major von Meyerinck that I was able to include the historical leaf here as well as the king's drawing.

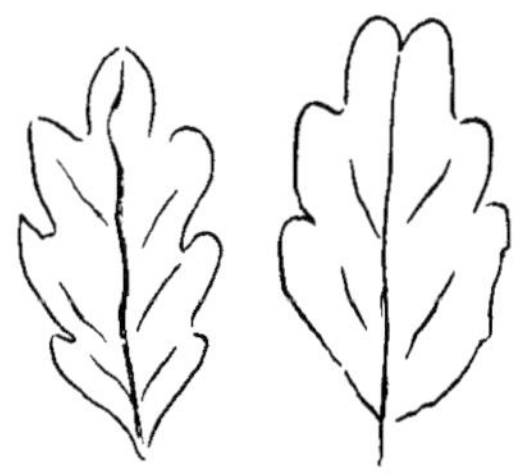

Fig. 18

Like the shape, the color of the oak's leaves changes, too. The spring, the Johannis-shoot [June], and especially the late fall, show differences and a magnificence of coloration which only a very unfocused viewer can overlook.

In nearly every forest there are oaks which shoot in copper-brown color, there are bright-red and very light green Johanni-shoots, and in the fall there are shades from dark brownish yellow up to very light lemon yellow which I have admired in *Quercus robur* more than once.

However, the **fall color** of these leaves is not sufficient for ambitious forest enhancers, and therefore they believe they cannot do without Red Oak [*Quercus rubra*]. But the **fall yellow** of our oaks complements especially the conifers in quite an excellent way. It offers a wonderfully appropriate supplement to the color of the pine and spruce, while the pale brownish yellow steps forwards a little bit in relation to the dull green of the pine, but steps backwards a bit in relation to the bright green spruce.

Young oaks and some of the older ones keep their leaves all winter long.

The horticultural nurseries abound with **oak varieties,** many of which are from German origin. Still, their number is growing. The Yellow Oaks *(Quercus robur 'Concordia')* probably stem from the Trebnitz District (out of Koschnoewe). The Silver Oaks *(Q. robur fol. arg. pictis)* [*'Argenteopicta'*] have been distributed out of the park of the Counts of Goertz at Schlitz near Fulda. The Blood Oaks *(Quercus robur fol. Atropurpureis)* have been found by Bechstein in the Lauhaer Holz of the Dukedom of Gotha at the beginning of this century. The progenitrix form of the Pyramid Oak *(Quercus robur fastigiata)* has also been known by Bechstein. In 1876, when Petzold described it, it was said to be 100 feet high, and its age was estimated to be 280 years. It stands near Babenhausen, between Diebau and Aschaffenburg in the Grand Duchy of Hessen. It is owned by the community of Harreshausen. The present measurements, given to me by Wilbrand, are the following: diameter at breast height 1m, height 24.75 m, branches beginning at 10 m height. Although hit by lightning 20 years ago, the tree is still alive.

I omit numerous other varieties because there are too many of them. The Count of Arnim Nursery in Muskau Ober-Lausitz names 63 varieties of the two native oak species in its price list. Here in Postel I found 13 wild, very different forms, some very interesting.

Beech.
[*Fagus sylvatica* (European Beech)]

Beeches are not easy to paint. Their representation is so difficult for the brush that perhaps only the master's hand will succeed, for the regular fan-like structure of its branches very easily looks stiff in the painting, especially in the tops of younger beeches, whose upper branches strive rather straight upwards.

The delicateness of the outer tips of the branches, which decorate themselves with marvelous foliage in summer and magnificent buds in winter, nevertheless does not give the impression of stiffness in nature, at least at that time when the quickly lengthening shoots do not yet have enough strength to carry themselves, and when they are hanging, bent down in a beautiful bow. At that time the young beech probably gives the most lovely sight our forest can offer. One cannot recognize from these little trunks that they are bound to grow up to an even higher level of beauty, to stands of the most solemn and most superb magnificence like no other tree species can show. With oaks, it can happen that one cannot see the forest because of trees, for they want to be appreciated particularly and individually, each one in its character. Quite different the beeches: building themselves up in strict form, each one like the other as a slim mighty column in regular distance from each other, they put their tops together to a high dome which they form above delicate branching with well-arranged foliage. The solemn impression is further increased by the lack of any addition, which might disturb the character of the whole. Only Waldmeister [*Galium odoratum*], Sauerklee [*Oxalis acetosella*] and delicate grasses break timidly through the even-spread blanket of fallen leaves, and only rarely a light beam in its unbroken power reaches the ground. The long, well-rounded stems, comparable to stone columns, are dressed into a rather cold gray, but decorated by the delicate cover of lichens and mosses; its shade is excellently fitting in order to show the colors of the foliage to advantage, to make them look rather warm by the contrast, from the first emerald-green of the May-shoots up to the incomparable golden-brown of the fall, up to the red-brown with which at last the dried leaves cover the ground.

Can one imagine, however, that one could end up, with the help of fault-finding and puzzling, and this from a famous source, but anyway only in the study room, in totally misjudging our beech and thoroughly scolding it! The famous aesthetician Vischer says about the beech: "The stiff branches, only in the center bowed a little bit downwards, stick out in a cutting scratching line; the serrated, wide, elliptical leaf sits on a short stem, alternately opposite, and does not play much in the wind. The crown has no large, individual limbs, but rather forms a single shape in outline. From the trunk, the hardness of the wood can be recognized, stern strength is its expression, which therefore causes a well-shaped, healthy and competent, but dour mood."

So far Vischer, and another aesthetician, Bratanek, who is also not without merit, reprints this judgement not only merrily and cheerfully, the same way he reprints another's remark that the nightingale only sings at night, at another place, but he even feels obliged to add, that "in the beech we meet the character of ruthless energy of the male, which does not ask much about whether its plain manner is liked, but enforces to all sides its hard and stinging manner 'in spite of everything.'"

Poor beech! You are said to touch us violently and stingingly, you are said to create a dour mood! We know you better! We know to appreciate your delicateness when we find you filling up with tender mediation every gap in the stand between the rough trunks of old oaks, when we look at you spreading your arms protectively with motherly care above your offsprings, when our eyes feast on your colors with which your swelling buds greet the first warm sun beam, marvelously unfolded into lashed leaves, in order to dress you three times newly into a magnificent garment for spring, summer and fall. When you finally give your foliage to winter's frost as a sacrifice, you make it promise to you new garments. You yourself weave the ground out of delicate tips of branches, covered regularly with magnificent buds, so that the hoarfrost can embroider into it glimmering patterns, decorated with precious stones.

Who can ever forget the marvelous sight of an old beech stand decorated with hoarfrost of the most wonderful beauty, or of a garden group carrying a fresh hanging of snow, when a sunny winter morning has uncovered to him such splendor even just once.

And one is allowed to look at this scene without disturbing second thoughts, for the beech, being truly a tree of the low mountain ranges, is able to carry an adornment that would bury other tree species under its burden.

Not all the aestheticians judge as unjustly as the two mentioned above. Already Gilpin accepts some good things of the beech, although he puts too much emphasis onto the picturesque. His eye has feasted too often on the incomparable play of colors of the forest, when the oak is still green, but the beech wears already fall colors, so he could not help but be its friend. Even more benevolently the more recent authors express themselves: Berthold, Masius, and most favorably Rossmaessler and Jaeger. I also have to agree with Rossmaessler insofar that he, contrary to Bratanek's opinion, sees embodied the female type in the beech and the male in the oak. If, however, both of them, Rossmaessler and Jaeger, declare that "the beech is definitely our most beautiful German tree," and the beech forest is the most beautiful that the German plant world brings up in the landscape, then I cannot necessarily agree with such judgement. According to my opinion (and here I can refer to Petzold), the beech **stand** is owed such special praise only where its "sublime unity" (using Jaeger's term) is protected against monotony by changes in the relief; as an **individual tree** I would prefer to treat the beech as equal to the oak only when an uncommonly old age has given it magnificent form.

It is not possible to prove it historically, but there is every reason to believe that the immense halls of untouched mature beech stands with their columns, with their architecturally regular branch structure carrying the roof of leaves, have inspired the emergence of the Gothic building style. The beech forest is more suitable than any other forest to make one feel reverent, because the rising trunks guide the eye upwards, while distracting ground adornment, which might keep the eye busy down there, has been avoided—could it be intentional?

Of the numerous **varieties** of the beech, especially the Blood Beech [*Fagus sylvatica v. purpurea*] is remarkable. The mother trunk of all Blood Beeches has been described repeatedly. I owe the following data to the kindness of the Princely Head Forester Spannau, under whose care the tree is carefully tended in the district of Oberspier (in the Principality of Schwarzburg-Sondershausen): The trunk is 27 m high and has a diameter of 100 cm at breast height. The crown covers an area of 380 square meters. When planted, its seeds produce 50 to 60 percent Blood Beeches, a relatively surprising high amount, because pollination by nearby common beeches can certainly be assumed. During the regeneration of the surrounding stand, its neighbors have been maintained as a canopy together with the Blood Beech, in order to avoid the dangers of isolation.

In May, 1898, I was able to determine that the Blood Beeches do have a high forest aesthetic value, when State Chief Forest Ranger von Strauch showed me the Blood Beeches in the forest district of Ettersburg, near Weimar. Being surrounded by May-green trees, those trees presented a wonderful contrast. But the forest picture would become too disturbed if one applied this conspicuous tree species too often. It should be planted only at places which are meant to be extraordinarily decorated. For example, I made the suggestion to plant four Blood Beeches in the Kranichstein Forest at the Hartig-Memorial.

Here in Postel I found a beech about 50 years old whose lower branches hang heavily downwards. If carefully cut free it would promise to become very beautiful.

The Muskau nurseries name 34 varieties on their list, nearly all of which deserve some attention.

Hornbeam.
[*Carpinus betulus*]

Nomen-omen [the name is an omen]. In a grove-type thin

stand, the multi-formed hornbeams grow into very beautiful forms. Their gnarled trunk is characterized by often highly distinguished strong ribs which make it appear picturesque. The branching is sometimes uprising, sometimes more or less level or even hanging down. The branches carry an abundance of deep-green foliage that separates itself into beautiful masses. Thus this tree species is very popular with landscape gardeners. The gardeners appreciate **varieties** with differing leaves, too. Especially the oak-leaf variety is interesting for every nature-lover, because the leaves show the most different forms all on one and the same branch.

However, the beautiful **fall colors,** which some hornbeams usually take on, are not appreciated enough. The "red-leaf" variety available in nurseries does not possess red summer leaves, only the May-shoots show some reddish leaves, but in fall it shows a magnificent copper red brown, which stands out very beautifully against the golden yellow of the ordinary hornbeam fall leaves. *C. betulus carpinizza* Hort. dresses itself even more magnificently in fall. No American oak can be more beautiful!

Ash.
[*Fraxinus excelsior* (European Ash)]

Being actually a marvelous tree, the ash is especially suited to decorate entire stands by only a moderate sprinkling, because its branching and its pinnate foliage are so much different from all the other native tree species. Even if there is a lack of good ash soil, it will be possible every now and then to enable the growing of some ashes by deep loosening of the soil.

If it is desired to bring some change into larger plantings of ashes (especially in allees), it will be possible to use the seedlings of garden varieties with the necessary caution. For example the one-leaf variety, growing as lushly as the common ash, showed at my place a very interesting group of off-spring with totally pinnate, partly pinnate and even non-pinnate forms.

Maple.
[*Acer pseudoplatanus* (Sycamore Maple)]
[*A. campestre* (Hedge Maple) *A. platanoides* (Norway Maple)]

On May 14 of this year, I had an experience with a group of maples, so to speak: I followed the kind invitation of Dr. Carl Bolle and visited the Isle of Scharfenberg and the nursery of Tegel under his charming and proficient guidance, according to an old wish of mine and with the special intention to study the forestry value of foreign tree species in this classic experimental area. I was especially interested in this "magnificent group of more than 100-year-old Red Oaks, stemming without doubt from the Burgsdorff days of glory" which Dr. Bolle had described. Looking for these in the mixture of larch trees and other remnants of those plantings, I discovered some magnificent trees with full foliage, which I admired and recognized immediately by their smooth bark as being the desired American Red Oaks. With this sight I was already nearly determined to warmly recommend the culture of Red Oaks for forest enhancement—when, stepping closer, I recognized the same oaks as being three Norway Maples and one Hedge Maple! The Red Oaks I had looked for were standing near by—they were very mighty trunks, but their foliage not yet fully developed and the branching not particularly picturesque.

All three maple species have in common the beauty of the foliage, and the advantage of having early green leaves and beautiful fall colors. The most splendid development of foliage is shown by the **Sycamore Maple,** which has light bark flaking off similar to the Plane tree [*Platanus acerifolia*].

Norway Maple is effective by the vividness of its colors: by golden blossoms and lush green foliage in spring, by light yellow and sometimes carmine leaves in fall.

Hedge Maple is, as a shrub as well as a tree, one of the aesthetically most valuable elements of the forest. Its delicately carved foliage takes on a gleaming yellow color in fall. Older trees grow into most picturesque oak-type forms.

Gardeners cultivate innumerable maple varieties, but only few of them are usable for forestry.

Schwedler Maple [*A. platanoides* 'Schwedleri'], a variety with blood-red leaves in spring, named after Court Gardener Schwedler, who found it among seedlings in Upper Silesia, is only fitting along roads, because it is too conspicuous; but by selection and cultivation of seedlings, one can grow less contrasting foliage. The same is true for the magnificent Sycamore Maple, which is red on the reverse of the leaves. Its seedlings show more than 20 different shades here on a seeding bed. Out of all three maple species, one can find trees with beautiful red seed-wings; preferably from those trees seeds should be used for planting.

Elm.
[*Ulmus glabra* (Wych Elm) *U. laevis* (European White Elm)] [*U. campestris (U. carpinifolia)* (Smooth-leaved Elm)]

Common to all three of the elms is their magnificent crown structure and abundant foliage. In this respect the most beautiful I know are the Wych Elms on the top of the Zobten Mountain. Gilpin has already written that no other tree species is more suitable to take on great amounts of light. This praise is understandable, because he probably has known especially, if not exclusively, Wych Elm. All three species delight us by their early blooming. Many Smooth-leaf Elms and some *Ulmus laevis* take on a wonderful red-brown, partly carmine color of the leaves in fall. *Ulmus laevis* covers itself with an abundance of brownish seeds nearly every year after blooming and in this state the tree is unsightly in the foreground. But it offers a favorable background for forward-stepping colors. Unfortunately, I had just too many opportunities during the April of recent years to look out of room No. 17 of the Parliament Building over the treetops of the Thiergarten [Berlin Zoo] up to the Brandenburg Gate. However, I was delighted to see how beautifully the blooming Norway Maples stand out from the after-bloom of *Ulmus laevis,* and the other way around, how good the pale brownish tops look behind the light green-yellow ones. This observation was another proof to me that every color is beautiful if it is used in the right combination.

The nursery gardeners cultivate numerous varieties, some quite attractive; however, I never see fall colors in the elm's leaves in the gardens as beautiful as they can be found free in nature.

Wild Fruit Trees
[*Malus sylvestris* (Wild Apple Tree)]
[*Prunus avium* (Sweet Cherry) *Pyrus communis* (Common Pear)]

The wild fruit trees, namely **Wild Apple Tree, Common Pear** and **Sweet Cherry,** decorate the spring by rich splendor of flowers, the fall by colorful shades of the leaves. The pear tree especially is characterized by a change of colors. It dresses white in spring, green in summer, red in fall, black in winter. Solitarily standing pear trees give a melancholic impression in winter by their color and the delicate tips of the branches, reminding one of tangled hair. The old Wild Apple Tree, however, is standing there defiantly, "bristly like a wild boar" as Burckhardt writes very typically. The Sweet Cherry appears merry; shooting up straight, it dresses its trunk with smooth bark, and it decorates its top with shiny foliage reminding one of laurel leaves.

3. The Conifers.

Pine.
[*Pinus sylvestris* (Common Pine, Scots Pine)]

Out of the number of coniferous trees I put first the pine, preferring it because it is certainly the most interesting coniferous tree, if not the most beautiful.

I nearly feel like singing its praise in exuberance, but as modest as it is, it would not be in its sense at all, if someone emphasized it excessively. Nobody has done that, however; quite the opposite, rather much undeserved criticism has been its fate. The pines growing on soil of the fourth or fifth class of value, particularly often accompanying railway lines, do not have much cheerfulness for the view of the weary traveler, I have to admit, but even under more favorable conditions they usually do not appeal very much to the modern civilized human, who lacks time and understanding to study the details. In respect to deeper conception of the entity, however, he lacks the survey and the calmness of the mind, and then the spoiled tourist is soon ready with his negative judgement. But it is different for us foresters, for we look with a benevolent and trained eye.

This benevolence is mutually based, for even when we approach the tree, the pine stretches its arms up to us in a friendly and inviting manner. At the **forest's edge** it lowers its branches down towards us, protecting the interior against wind and sun. The branches curve downward and then rise again at the tip in an S-shaped swing, friendly like the branches of the hospitable lime tree. Someone who steps up to us like that, with a hollow hand, cannot be an enemy. The convex-formed, pointed up jutting is hostile, therefore the pine avoids and moderates such forms as much as it can. The rounded trunk is dressed with elastic bark-scales of warm and friendly color, peacefully populated by unpretentious lichens. The long needles are soft and pliable, their bundles uniting to the shape of a swelling bolster.

One may distinguish three quite different forms of appearance of the pine: young pines, old pines grown up in a closed stand, and solitary old pines; these differ more from each other than, for example, the lime tree from the elm, or the fir from the spruce. Who would be able to recognize the ideal of symmetric regularity of the first decade in the later forms (be it a tall, slender trunk with umbrella-like top, or a deep and irregularly branched, open grown tree) if he had not been accustomed to know this tree since childhood.

In its youth, pine shows the typical coniferous character in the perfect regularity of its structure, the symmetry of its branching and the conical shape of the crown. Now there is no doubt about it—beauty based on symmetry is one step lower in range than the free but balanced grouping of the deciduous trees; where, however, the pine's stiff form gains a charm of its own by delicateness, as long as they are still young and small, there coniferous trees are very well able to struggle with the deciduous trees for the first prize. Who might appreciate the Christmas tree less than the May-birch! However, not often is the pine chosen for a Christmas tree over the more vividly colored spruce, more delicate by the amount of tributary branches; but our friend knows how to comfort itself over this neglect. If **we** do not decorate it with candles, it itself will set candles upon it, long before it is getting ready for the holidays, for it knows pretty well that it has to manage its means economically and prudently if it wants to continue to exist on its poor site, or to exist with honor. So it has already formed the young shoot in fall, enclosed in a warm bud and ready for the upcoming May. Already every pair of needles knows the place of the delicate spiral, where it is supposed to break through and to remain for thirty months, for there the bud scale is already waiting for it in order to protect its first period of youth. That bud scale, only meant to be for protection, not dressed in the material-transforming green, provides the young, upright shoot with shimmering bright coloration. This is present just at that time of year when all the plantations are ready and the forester has the leisure once to take all the family, old and young, with him into the forest. Then the young folks jump from one young tree to the other one: "This is my Christmas tree, this is yours, you light that one, and I this one," and now they start working, a little stick in the hand for lighting, without flame, however, but with vivid imagination and rather loud merriment, always from one pine to the other one. This is then a family celebration, a spring celebration, worthy to be put beside the "Herbstling " ["Autumnal Fruit"] and "Eiszaepfel" ["Icicle"] of good old Claudius, the celebration program of which you can read in the *Wandsbeker Bote, Part Two.* My children have invented it themselves, but they are not going to dispute Rossmaessler's right to priority, with which we already find the germ of the same thought. The pine forest incidentally is a true children's forest in winter, too; it is warm, windless, dry and rich with a thousand small things which children love. There are the old cones as a nice toy for the youngest, then the coral moss is appreciated, as is the magnificence of the *Hypnum* species. Back home then, in the warm room, out of the collected materials, the small ladybird is crawling (the Silesians say Sommerkaelbel); this then is an event, this visit in the room in winter; but outdoors there are still quite different things to admire. There is the friendly goldfinch that can be looked at so nicely from close by; there are the busy titmice, the hammering woodpecker, and under the *Juniperus* shrub, the hare is sitting and thinking. Then it has to pay attention to the dachshund that likes to accompany the children, otherwise it is going to chase it and to hunt without end. As far as I am concerned I do not want to behave like the ill-mannered dachshund, do not intend to be misled away from my path forever by the warm trace of the hare, but would like to go my way as systematically and pedantically as the pine builds itself up, even beyond the **polewood age** up to the mature stand! The delicateness of its youth is now already long gone, but still it knows how not to appear stiff, although it is stiff. For the shoots are not as long as they were in the beginning, therefore the small branches can condense the crown considerably and can round charitably its beautiful egg-like shape. Now the pine does not stand anymore in contrast to the appearance of deciduous trees, as it did during its early youth, but adapts to them harmoniously, therefore it can often be aesthetically beneficial in combination with them, but never damaging; so it also fits into deciduous stands as single or group-wise sprinkling, as the other way around, deciduous trees intruding into pine stands are favorable in every season. How this is to be understood one can realize easily by comparison with the spruce. This species, with its pointed top and its pointed branches, its regular story-like structure, its abundance of needles and its dark color, is contrary to the deciduous tree in every respect; that is why it can be more effective than the pine at the right place, but can spoil a lot, too, at the wrong place.

Wherever the pine is allowed to grow freely, the rounded forms of its branching disunite more and more into single separate groups, and then with every year its approach to the deciduous character becomes more perfect. It is very entertaining to regard trunks like these individually in respect to how with each one, the transformation has taken place or is beginning. With close observation one can find that especially the tense equilibrium into which the development-sequence of the pine is settled becomes the cause for its development into more free forms. Especially, because all the lateral buds, grouped around the center bud of each year's shoot, spring off from it at one height and into one direction, and if then only one of them is favored by a coincidental circumstance (a superior upright position) in comparison to its peers, it will soon gain a great lead over them, which keep themselves in balance all together, just as in a democratically-leveled republic a dictator may swing himself up very easily. One like that usually enlarges his field of power by force, and so does our pine shoot, too. Enjoying more light than its brothers from the same year, it does not only outgrow them, but also the next higher, younger whorl of branches, and so it conquers more and more light space for itself, until it finds an equally strong opponent in another branch, also favored by a special position. This is the regular course. The intervention of insects and fungi, crushing and breaking by snow, the attacks of deer and cattle, and every now and then human action, too, usually interrupts this regular course and gives cause to the most interesting, often very beautiful formations, sometimes more imaginative than beautiful. Mostly several such influences are effective, one after another, in order to give character to a tree. The plant world in its entirety and each member in particular are more interesting to us than inorganic nature, just because it has its fates like humans have, to which one can so easily lend the appearance as if "the plant **experiences,** too, what it lives. How quite naturally this lending takes place, daily experience shows. One hopes with plants, one looks at them as if they had a feeling of their power, one feels something like respect for this old man of the forest whom so many generations of living beings passed, one pities the fruit tree destroyed by frost, the oak killed by lightning, as if their fate was tragic." Due to the delicateness of its branching and the lightness of its needle bundles, the pine, however, will never give the marvelous, the **tragic** impression like the oak, although it bears the traces of former experiences more lastingly than the oak, because it lacks the ability to re-cover bald areas of trunk and branches again by epicormic shoots from dormant buds (like deciduous trees do). Those traces, however, often are not blurred, but veiled, for every pine branch possesses the ability to send shoots into the direction of the light, when the light gains access from a side that was blocked before; often in a direction opposite to the one the branch started its growth into and maybe still continues it. So the round forms of its crown come into being, which make such a friendly impression to us, because we only look at the entire form, while the details of the delicate branching, on which it is based, only attract attention upon close examination. (Quite different from the magnificently laid-out oak, with which just the crookedness of some branches gives a strong and energetic impression.)

The one-side branched pines, deprived **very** late of their peers, will, however, stay melancholy forever. Nevertheless, we do not need to look at **all** conifers, like Rossmaessler does, as **"lonely** and as **mourning strangers,** for since the period of coal formation is over, the friends from those other plant species, who used to mix their crowns among the conifers, have left them, and they feel it nearly like a sad privilege now to reign by themselves where they liked to share the power with relatives in earlier days."

Trunks that were dependent on themselves from early on and had to defend themselves against several attacks appear to be more beautiful than those which were released only long after they finished vertical growth. To them, the scars are a sign of struggle, a decoration. There we see withered tops, soaked with resin, keep for decades the memory of a tough struggle against the Blasenrost [blister rust] or a quick succumbing to innumerable masses of the Kieferneule [pine noctuid, *Panolia flammea]*. Even 50 years later, on such a trunk dry branches still show how it used to have its crown attached low and spread out, before it formed its current higher crown, forced by recent growth. The diversity is increased by the individual character inherent in our tree species, and by the diversity of the site. High-rising and wide-spreading growth, spiral growth of the trunk or of branches or the whole tree, slender or crooked shape of trunk, a tendency of the branches or the stems or both for hanging downward—all these are appearances which are determined as much by inherited qualities of the single pine varieties as by the site. These diversities show to advantage, especially with actual disturbances.

Like we humans in general do not show the advantages and faults of our character as reliably in everyday life as in times of hardship or extraordinary happiness, the pine grows after a fashion on medium site qualities; but on ortstein, or on wet sites, or on the poorest sand, it is as likely to diverge from the everyday shape as on rich humus soil. Under these conditions the characters of the varieties and the single individuals show off very soon. First I would like to recall the strange pine tops near Eberswalde, which, grown in cypress form by the "forest gardener," interested Ratzeburg so much that he decorated the title of his publication *Forest Insects* with the artistically interpreted picture. This picture probably is known by every professional colleague; maybe less well known, however, is the accompanying text, which I am going to quote here because of the parts emphasized (by me), out of which it is to be understood that Ratzeburg regarded the appearance as significant, especially from the aesthetic point of view. He writes:

"Here in Neustadt, directly behind the shooting stand, there are about a dozen old pines, which are not only very strange, because they have withstood the insect damage for an unusually long period of time and got quite a different growth by it, but also because **all of them stand on the height, and stand out against the horizon excellently, rising above the deciduous trees.** Some have the most striking similarity with cypresses, others with groomed Yew [*Taxus*] trees which used to be fashionable in art gardens, and, not inappropriately, Mr. Hylesinus has been called the *hortulani naturae famulus* [the master of nature's gardeners] by Linne. Our genius Roesel, therefore, grew to like them so much that he provided them for the book *Forest Insects,* presented in a nice picture."

Even better known are the changes in growth caused by shearing in the allee of pines near Darmstadt, the so-called "Scheppe Allee," but I do not wish to stay with the abnormalities here. I would prefer to recall those marvelous trunks of the Liepe Forest District, which were held over by Grunert at his time on previous old oak areas in the areas of Maienpfuhl and Breitesenn; these pines are centuries old now, released long ago, with mighty wide-spread picturesque copper-red branches and gigantic trunks of the same color. I do not know these pines by personal experience, however. The most beautiful pines known to me were shown to me by Kessler in the Rauen Mountains.

The pines of my former apprenticeship district, the Royal Head forester's District Katholisch-Hammer, are characterized by the forming of wonderfully beautiful clusters. We have accompanied the pine up to now while it takes on forms of deciduous trees, unfolding from rigid coniferous character to ever more free forming, so this occurrence can be seen in Katholisch-Hammer, too, of course, but in less magnificent scale than on soil of the Mark Brandenburg.

The shape which pine achieves in a totally open situation, free from youth on, is not the most beautiful one it can attain, in my opinion. It rises to its greatest magnificence **if it grows up totally according to its nature.** Where a small gap in the stand gets sufficient light due to windthrow or another cause and where the soil is just in the right condition, there we see the pine seed in and germinate very abundantly and grow up to a considerable height more densely than we can ever achieve or at least maintain in our artificial sowings. In this position, then, only the most worthy trunks out of the great number of competitors maintain themselves, in which the tendency exists to withstand the earth's gravity, rising geotropically straight up, until they reach the height the site permits them to get.

During this growth period, one century will pass. Now they begin to spread their branches horizontally. Densely needled, they stretch their branching towards the light; not seeking any more to use its material-transforming power for the production of great masses of wood, they now refine the wood already formed until it will be equal to the value of oak wood, each year widening the red heartwood. Its delightful sight, however, one cannot enjoy while the tree is still alive; in the meantime one may admire the beautiful bark. The structure of the lower part of the trunk is quite different in form and color from the upper part, giving an architectural structure to the long trunk without branches and knots. On the lower parts it is brown-red, picturesquely rugged and deeply fissured into rough scales, while on the upper parts, it is vividly colored, delicately furrowed, flaking off into fine layers. The top and the branches are dressed more and more lightly, wonderfully standing out against the deep, somber, dark green of the needles. Grown up to such a condition, the pine stands without peers. At least in our vegetation we do not have anything similar, and that is why it easily feels lonely, and gives a melancholy expression when standing solitarily, but if several of them, nearly 300 years old, grown up out of one cluster and united into a loose group, stay close to each other, or if they tower like an umbrella above another species, especially the beech, then they are of the highest splendor. Quite a masterpiece of shaping Mother Nature.

Whoever has read Rueckert's "*Nal and Damanjanti*" will remember with joy the 14th song when

> *Damajanti, deeply sad,*
> *Looking for her spouse, used to pain,*
> *Found wandering in the center of the wood*

The tree with the name Free-of-Sorrows
The wonderful one, free of sorrows,
The sad one started to caress:
Happy tree in the wood's center
Rising above like a king,
Decorated with many crowns,
Caught in no sorrow!
I was caught by a heavy sorrow;
Oh Free-of-Sorrows! Make me free of sorrows.

If we were to choose a friend among the trees of the forest in order to talk to it of joy and sorrows, to whom else could we want to go than to such an old pine. It is standing, too, now "free of sorrows" after some youthful yearning, high above us, but it is whispering down in melodic language.

But I have promised not to become effusive if possible; that looks worse in prose than in verses of Rueckert anyway. So now I am going to present as an expiation for the just completed offense a rather dry contemplation out of the applied teaching of colors. The pine's color is not a reflecting one, as long as illumination does not give a temporary shimmer to it, and therefore one can easily misjudge its value whenever one starts to compare and to criticize.

Especially a solitary young pine among spruces looks like a Cinderella, but already the old Gilpin, whom I had the opportunity to quote so often with the oak and the beech, has pointed out how no color actually is beautiful or unsightly. It all depends on whether it fits at its place and whether it has a good effect in the combination in which it appears. The pine needle has a modest backward-stepping color; it presents a very excellent background, delicately shaded in itself, for every large scale landscape scene, and in particular, it makes the cheerful spring and fall garments of its most faithful companions on poor sites, the birch and the aspen, show to their highest advantage: the other conifers,too, (spruce, fir, larch) look excellent when distributed among pines, and sometimes form a very beautiful combination with them. This does not at all contradict my earlier comments, for there is a great difference whether somebody is dressed more festively and mingles among a simple party, or whether the same person is supposed to sit at the table in a housecoat together with people all dressed in white collars. The latter one is very uncomfortable. For the same reason, **the pine does not fit into the garden,** at least as long as it is still very young. When reaching pole size, this is another matter already, for its trunk gains a marvelous redness by contrast with a well-groomed garden lawn; therefore the pines in the villa gardens of the suburbs of Berlin look quite good, particularly because their rounded tops in themselves present a well-effective contrast to the narrow-angled roofs and small towers of the fashionable villas.

In relation to its wide distribution and its numerous occurrences, the pine has developed only a few **varieties.**

Beissner distinguishes 24 varieties, which, while not very interesting to the gardener, are very interesting to the forester. No two mature pines look alike, unless a dense stand determines their form. Figure 19 shows a "weeping pine," which has maintained itself under the light canopy of a mature stand. Recently it has been cut free at one side and will be able now to develop in a very nice way. **The needles** do not show an obvious difference, although the **blossoms do.** The form with **red** male blossoms, which is rare in Silesia, deserves ample consideration at wood edges. Once I composed a bunch of pine branches with strawberry, sulfur-yellow, and orange colored blossoms. No decorative shrub can be more beautiful than this bunch was.

I do not yet know whether the red blossoms pass on by seed, and therefore I cultivated the red variety by grafting. Koch thinks that the reddish filaments occur more often in the Province of Prussia and farther east.

The grafting of conifers in the greenhouse is well known and easy. Pines are said to be able to be grafted outdoors, too; Head

Fig.19

Forester Schmittspahn describes a method to graft pines, often applied by the Grand Ducal Forest Master Heyer, as follows:

"Around Johanni [June], the leader, formed in May of the same year, of a common pine, being the basis for grafting, is cut off in a diagonal way, and another leader of the same age and diameter, from the pine which is to be reproduced, is shaped and fit to it. After the wild part and the scion have been put together according to the rules of the craft, a bandage is put on, which was made out of raw sheep wool according to Heyer's instruction; it is wrapped around the grafting place and tied with a woolen string. This string has to be loosened a little, as soon as the stem starts to grow. One can also split the leader of the wild part and set into the gap the May-shoot of the same diameter, sharpened on two sides."

Common Spruce [Norway Spruce]. [Picea abies (P. excelsa)]

The spruce is a solemn tree. Our earliest happy childhood memories are connected to the **Christmas tree.** When a village or town gets ready for a festive reception, then spruce poles, connected with garlands of leaves, decorate the narrow streets. At a sad occasion somber spruce green together with *Thuja* and palm branches decorate the rooms. The common habit which prefers the spruce in this way is not only based on the advantage of durability, which is a feature of the needle garment of the spruce, but even more on its beauty.

The symmetrical delicate branching structure and the lush dark green color of the rich needle garment are the main advantages of all spruces. With old age the beautiful contrast of branch-forming is added, when the thin stems hang down from the stronger branches like a curtain. While the nice red blossoms mostly hang too high for easy contemplation, the large cones are more obvious.

When spruces with a disposition for characteristic growth must suffer rather a lot of damage, from browsing by deer and cattle up to top-breaking and lightning, then those wonderful, highly picturesque forms develop, which have been drawn so often by Calame and Dore—the often admired "weather firs."

Beissner lists 61 spruce **varieties.** The spruce was fortunate to be the first to have the multitude of its forms appreciated in a special publication. Its author, Dr. C. Schroeter, distinguishes varieties, lusus [sports], and forms.

The **varieties** contain those spruces which appear in a greater number and distribution, related by several hereditary markings. **Lusus** [sports] are characterized as well by hereditary markings, but occur only individually and are usually not connected by transitions with the adjacent plants of the same species. **Forms** are distinguished by markings which are not hereditary. The varieties of the spruce are distinguished from each other by the shape of the cone scales. They are aesthetically equal. The lusus distinguish themselves from each other, furthermore, by branching, needles, bark, shape and color of the cone. I call the **Hanging Spruce** particularly beautiful (also called **Hazel Spruce**—*Picea excelsa Lk. lusus viminalis*). Its first order branches are nearly horizontal. From these, very many second order branches hang down limply up to three m, in single cases even up to six m. This lusus is said to provide the best wood for sounding-boards.

Most opposite to the Hanging Spruce is the **Snake Spruce** *(Picea abies 'Virgata'),* which shoots only very few first order branches, and which themselves build few or no side branches at all. This lusus is not beautiful, but it offers an interesting variation in less distinctive specimens. **Column Spruces** *(Picea excelsa 'Columnaris')* with shortened, richly structured side branches, seem to be very beautiful, as far as the pictures at Schroeter and Conwentz allow to recognize. **Vertical Spruces** *(Picea excelsa 'Erecta')* turn their branches straight upward beginning from the ground. I know **Weeping Spruces** *(Picea excelsa pendula),* whose first order branches hang downward along the trunk, as young park trees. They are not pretty. It is still to be seen whether they are going to become more beautiful with age. The **dwarf lusus** *(lusus nana)* are similar to the variously shaped *Juniperus,* without reaching its beauty. It may be a matter of taste to decide whether the **red cone or green cone lusus** *(Picea excelsa F. erythrocarpa* and *Picea excelsa F. chlorocarpa)* are more beautiful, but it is important that each one of the two colors appears rather clearly. Undistinguished mixed sorts are less attractive.

The color of the needles is particularly very beautiful in the first weeks after growth begins. One can probably find bluish-appearing needles in every planting garden. More rarely, the golden or whitish lusus occur, which are also cultivated in gardens. When they happen to occur on his planting beds, the forester is not going to overlook them.

More important to the forester than the lusus are the **forms,** whose outer appearance is determined by site or other influences. Dr. Schroeter shows the **"pointed spruces"** of high mountain regions out of Graubunden [a Canton of Switzerland]. One can walk through quite similar stands in the Riesengebirge [Sudenten Mountains], above Krummhuebel on the way up to the top. Repeated snow breaking, lightning, landslides, cattle pasture and deer browsing often cause them to develop marvelous and highly picturesque shapes.

I remember two highly unusual spruces out of my own forest, which always drew my attention to them. When spruces were still rare in Postel, the deer used to exercise their antlers with particular preference at those little trunks. Thus prohibited in their vertical growth, some spruces developed their lower branches so lushly that they were lying heavily on the ground, forming root-like parts and raising their tips into new crowns, to which the same fate befell as had the main trunk. A spruce family struggling for existence in this way may spread far, and only the attentive nature lover will recognize that it is one single tree which has expanded to a whole group.

As variously shaped as the spruce may be, there is, however, nothing more monotonous than well laid-out spruce plantations and thickets of one age; for, being intolerant towards even its own species, in a closed stand it suppresses early any attempt of single small trunks to differ from the great majority's form. That is why one always should have a mixed stand. The spruce itself is very well suited to grow in mixtures. Under the light canopy of pine stands, it develops into a very decorative under-canopy or middle-canopy, which makes by contrast the reddish color of the upper pine trunks step forward vividly.

European Silver Fir.
[*Abies alba (A. pectinata)*]

As the beech relates to the oak, so in many ways the fir relates to the spruce. The **Silver Firs** have several advantages over the spruces: the wider, shiny needles decorated with nice lines on the underside, the powerfully upright cones, in old age the downward curving top, out of which sometimes, like Masius remarks accurately, "the branches grasp into the air."

The ease with which firs regenerate naturally and the remoteness of their mountain sites make the observer forget how much the fir stand also owes to human care. Under its "dark steel-blue shade" the hiker feels "shivery in the center of the forest; forest solitude in the green shaded, resinous-smelling hall, where the vegetation is by itself and, in its freshness, knows nothing of the sweat of the struggling human life."

Beissner distinguishes 13 varieties, which are characterized by special growth, color etc. I cannot describe these varieties as they are still unknown to me.

Common Larch.
[*Larix decidua (L. europaea)*]

Similar in shape to the spruces, the larch distinguishes itself from the other conifers by its delicate colors. Therefore it is very suitable for brightening up the forest scenes by a beautiful but not too strong contrast. For a great part of Germany the larch (as well as spruce and fir) is not a native tree species. The necessary care has not been used in its cultivation, and, as a result, this has had a bad effect, both economically and aesthetically. The eleven varieties and lusus mentioned by Beissner can hardly be of particular importance.

Common Yew.
[*Taxus baccata*]

The dark green of the yew needles, which is not brightened by a strong gleam, as with the fir, among it the magnificence of the red berries, the picturesque branching structure and the light-colored bark make this legendary tree one of the most valuable inhabitants of our forests. Where it still occurs, it should be taken care of and bred carefully. That would also be economically useful due to the high value of its timber. In the Prussian State Forests, protection of yews is required by law.

Gardeners grow very many varieties from cuttings whose sparse growth may have provoked a prejudice against the yew among foresters; but seedlings grow excellently on an adequate site. Nevertheless, I do not want to recommend the reintroduction of the yew in an area where it has already vanished, for where population and wildlife are not familiar with the poisonous branches, there its cultivation may be dangerous.

Arve
[*Pinus cembra*]

I have not yet had the opportunity to see the arve (Siberian Stone Pine) on its natural site, but I know the old arve tree on the Peacock Island [Berlin]. It gives an idea of what beauty it is capable. Also, its magnificent cones, surrounded by rich clusters of needles, which, together with edelweiss, I first held in my hands many years ago, have remained unforgettable to me. I quote the character of the high mountain arve out of Masius: "The originally whorled branching bows down in sturdy lines and then up again; every now and then the storm or the winter with its burden of ice may have torn a gap, every now and then long grey lichens hang down; but in the crown the bushy roof curves, impermeable to snow and water, while the roots tie up around rocks and blocks with a giant loop.

So the arve withstands the attack of the elements, and if nevertheless the crown or the trunk has been broken, then the strongest side branches will grow upwards as sort of new trunks and will position themselves sheltering around the missing leader."

4. The Soft Hardwoods

Linden
[*Tilia cordata* (Limetree, Littleleaf Linden)]
[*T. platyphylla* (Summer Linden)]

Who is not going to praise a limetree,
But is everybody going to plant it ? No.
It should be less elevated,
But planted more often!

So I compose freely after Lessing. The limetree, not the oak, is the proper German tree. It seems to be more ideal than the oak, because it addresses the soul by its richly given fragrance of the blossoms and because it is less useful. It reminds us less of everyday needs which are waiting for their satisfaction.

Greening early, offering beneficial shade and fragrance, inhabited by honey-bees, healing every wound, it has been the preferred tree for the center of the village from early on.

In the dense stand of the high forest it gains a beauty of a quite different type as a straight, vertical trunk, so that one can hardly recognize it.

I take an interesting dialogue on the limetree out of a critique of the first edition of this book. The critic is talking to a painter about the chapter "Teaching of Colors." There the painter picked up a limetree leaf and exclaimed: "Look at this leaf! What kind of work is this! How accurately this is painted! In spite of the delicate borderline it is nowhere exceeded. It is a very subtle color with which nature paints, while we only can smear coarse material on the canvas. By the way, nature, too, is terribly bilious green outdoors in the spring. Take a look once at the Siegesallee in the Thiergarten from this aspect in late May. The eye is frightened by these masses of verdigris and cannot stay on them with pleasure." Praise as well as criticism are very true here. The forester, who never cultivates the lime as a pure stand, will not give cause to the latter one, to criticism. Limetree alternating with oak looks very good in an allee in this area.

In Klein Commerowe in my neighborhood, there are limetree allees which stem from the times of the monastery, whose crossings and end points are marked by spruces. Also in this combination a critic would not need to fear from "masses of verdigris." More important than the many **varieties** and **races** named in horticultural catalogs are, to the forester as well as to the beekeeper, those which bloom particularly early or particularly late, because one can extend the wonderful pre-summer, the time of the limetree-fragrance, by planting them. Five weeks then may pass from the blooming of the earliest limes to the end of the blossom of the latest ones.

Poplar.
[*Populus nigra* (Black Poplar)]
[*P. X canescens* (Gray Poplar)]
[*P. alba* (Silver Poplar)]
[*P. X canadensis* (Canadian Poplar)]

Poplars usually are characterized by delicately-structured crowns, which allow light and wind to penetrate them. "Jumping lights" owe their existence to this fact. The quivering, shiny foliage throws the sun's reflection here and there, and on the ground the sun beams hurry here and there alternately with the shade. At the same time the trees are talkative. Their whispering tells us the direction of the wind, often as reliably as the observation of the clouds' movement. At the time of their blossoming, the long catkins blow beautifully in the wind. Every poplar species has its particular advantages.

The **Black Poplar** takes on much more picturesque shapes in old age than the Canadian Poplar does. Its light bark makes it appear very beautiful in the foreground in front of conifers in winter. Old female trees offer a very strange sight at the time of seed ripening, when large white bushes of wool cover the top, but which is not liked in residential areas because the seed wool penetrates everywhere and may become annoying. When the wind plays in the tops of the **Silver Poplars** a marvelous change of beautiful pictures is created. Marvelous, too, is the fall color of this tree species, which is not cultivated in a sufficient amount. Similar to the Silver Poplar, the **Gray Poplar** is characterized by a particularly picturesque growth.

The **Quaking Aspen** [*Populus tremuloides*] with its pendent catkins is one of the loveliest messengers of spring, and it embellishes late fall with its golden, sometimes even carmine leaf color.

Alder [European Alder].
[*Alnus glutinosa*]

The dark bark and the dark foliage of the alders present an excellent contrast to other softwoods, especially in late fall, when the alder leaves are still green, while other deciduous trees are already colored. I have never been able to observe here that the leaves of the alder take on a yellow color, as Kraft notices. In **winter** the crowns of old alders, covered by cones and catkins, take on a very beautiful brown-red color, which fits perfectly to the green of pines.

They brighten up the forest by attracting the vivid siskins. The King's Alder of the Spreewald, whose maintenance was ordered by King Friedrich Wilhelm IV, proves that the alder can grow up to a mighty trunk.

Birch.
[*Betula pendula (B. verrucosa)*]

Being light green in May, golden in fall, and decorated by

hoarfrost in winter, the birch always is dressed delicately and lovely, but in infinitely different ways. The **Ruchbirke** [Hairy Birch, *Betula pubescens*] has three advantages over the others. Its trunk dresses itself into an even more faultless white garment, its foliage offers delicious fragrance in spring, and keeps a lush green color for a longer time.

What has been said about the spring green with the limetree is even more valid for young birch foliage. Birch forests should always have aspens or oaks shooting out in brownish colors, and if possible, even conifers.

Whoever knows the Beresina-panorama by Fallat will have admired the magnificent colors which the **branching** of far away birches reflects in the evening sun. Pointed out by the artist, one is going to notice the same advantage in nature more often and will be pleased by it.

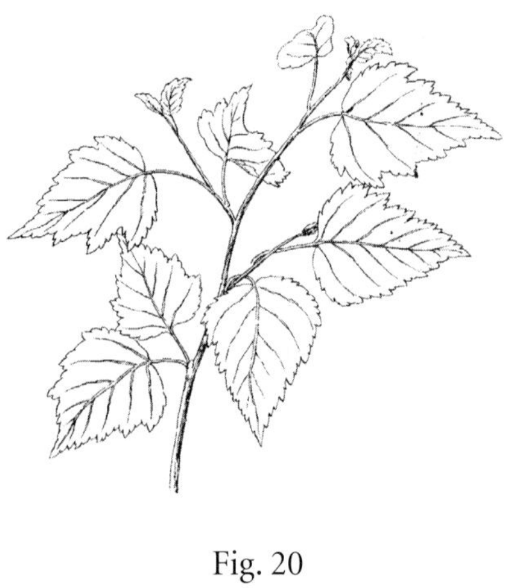

Fig. 20

Weeping Birches [*Betula verrucosa 'Youngii'*], with branches hanging far down, are found particularly often among *Betula verrucosa.* With calm weather, when their regularly distributed leaves are glistening in the sunshine, they remind one of an artificial fountain's falling drops. They occupy the eye and the imagination even more, when the wind or the storm are whirling around the swaying switches. It seems to be quite doubtful to me whether our dendrologists are right when they distinguish only two birches growing to tree size in the German forest; they look at the Hairy Birch as a variety of the pubescent birch, but I lack the time necessary for an independent study of this subject.

For the aesthetic importance of the distinction, I show here the branch of a young *Betula verrucosa* (Fig. 20), beside which the intensively smelling top of the branch of a Hairy Birch is represented in the natural downward position as a proof that Hanging birches occur with this species, too (Fig. 21).

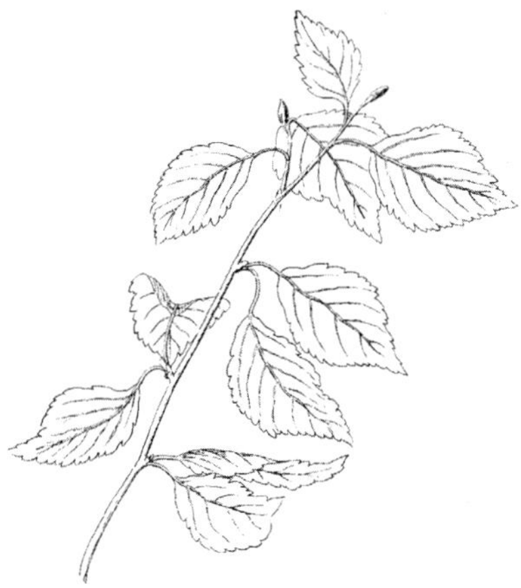

Fig. 21

In order to prevent inappropriate plantings, I would like to remark that the Hairy Birch needs more soil moisture for a good growth than does *Betula verrucosa.*

Willows.

For the willows, partly the same is true as I said in introducing the poplars.

Some species are standing in a close relationship with air and light like the poplars, therefore the poetic expression is very accurate when Rueckert called their delicate branching plumage.

Out of the great number of species I pick up some, which are better known to me, for discussion:

Five-men willow. The magnificent shiny foliage is similar to laurel, with large male catkins standing up excellently in a bright golden yellow. The silver shiny seed wool sticks to the female bushes until deep into the winter. The branches are nicely brown-red.

Silver Willow [*Salix alba*]. It grows into magnificent picturesque trees. Numerous varieties allow the most manifold use. The ones with leaves that are dark green on the surface and white on the underside show themselves especially beautiful in the wind's play. In winter the golden willows are particularly decorative, and are distinguished from each other by more or less yellow or reddish branches, and by hanging or upright growth.

Crack willow [*Salix fragilis*]. Bright green foliage, splendid growth.

Almond-leaved Willow [*Salix triandra*]. The trunk is peculiar; it peels off the bark in large pieces, like the Sycamore Maple. Those should be preferred whose foliage is colored shiny dark green on top, pale green (sea green) on the underside.

Bay Willow [*Salix petandra*]. The branches are blue-frosted, the catkins develop early.

Goat Willow [*Salix caprea*]. "With still un-foliated branches in March, it shines out between other trees due to the abundance of blossoms on its branches, covered with golden yellow catkins."

Shipyard willow. Dr. C. Bolle calls it "water crooked wood."

Eared Willow [*Salix aurita*]. The abundant blossoms have a particular pleasant fragrance.

Hemp Willow [*Salix viminalis*]. Among the many varieties, those are preferred which have yellow bark, and whose foliage is dark green on the upper surface and white below.

Creeping Willow [*Salix repens*]. Multiple forms with delicate foliage.

Purple Willow [*Salix purpurea*]. Beautiful purple catkins on male bushes.

Sorbus species.

The variously shaped genus of the *Sorbus* with its three main representatives, the Common Mountain Ash [*Sorbus aucuparia*], the Whitebeam [*Sorbus aria*] and the Elsbeere [*Sorbus terminalis*], to which the Swedish Whitebeam is added in the Northeast, are very suitable for decoration of the stands by their shape and color of foliage, early greening, rich fragrant splendor of blossoms, and especially by the bunches of berries.

The foliage of the Elsbeere, serrated similar to *Platanus,* is characterized by beautiful fall color. All species liven up the forest directly by inviting the birds to their crowns (thrushes, bullfinches, and others) and wildlife below their crowns. Whoever plants a *Sorbus* only here and there at road edges will soon have the pleasure that it invades as a friendly undergrowth into the stand, distributed by birds.

Bird Cherry.
[*Prunus padus*]

Like the willows, *Prunus padus,* too, offers a transition from the trees to the shrubs. Sometimes growing upright, but more often bowed down and irregularly branched, this tree species presents by its shape of growth alone an excellent supplement to the stand's scene in stiff alders, with which it often occurs. The young, lush green foliage and the white blossoms, nearly too fragrant, are so bright that they shine out of the inner stand nearly like they were self-illuminated.

5. The Shrubs.

When I visited the Nesigode Zoo I saw **European Red Elder** [*Sambucus racemosa*] flowering together with Bird Cherry. The yellow blossom, otherwise hardly visible, was strangely elevated by the contrast with the bright white bunches. The red berries, which appear in summer as a decoration of the forest to replace the splendor of blossoms, do not need such help.

Less colorful but much more magnificent in shape appears **Black Elder** [*Sambucus nigra*], with its large umbels of blossoms and black berries in front of its nicely arranged foliage. Quite a group of beautifully flowering shrubs are to be mentioned together with these.

Not in botanical respect, but regarding aesthetic points of view and the site, **Hazelnut** [*Corylus avellana*] and **European Cranberrybush** [*Viburnum opulus*], **Blackthorn** [*Prunus spinosa*] and **Pfaffenhuetchen** [*Euonymus europaeus*], **Common Buckthorn** [*Rhamnus cathartica*] and **Wild Rose** [*Rosa sp.*], **Hawthorne** [*Crataegus monogyna*] and **Schiessbeere** belong together. One cannot admire sufficiently the inexhaustible diversity with which nature mixes up here and separates there these shrub species, for example at the edge of a composite forest or along a major ditch. When one tries to figure out mathematically, with algebra, how many different combinations are possible, then one ends up at an incredibly high figure, and then one finds it understandable, too, that nature never needs to repeat itself. And how beautiful are the shrubs as individuals. The Hazelnut decorates itself with the flowing catkins as a first messenger of spring, then with the splendid rounded foliage. The European Cranberrybush with its umbels of blossoms is much more resplendent than the unfruitful garden variety; after all, it shows the bright light-red berries among fall-colored leaves. The Blackthorns [sloes] welcome spring with a white bridal veil. The *Euonymus* decorates the fall. Hawthorne and Dogwood offer with the lushest green all summer long the most beautiful background for the colored shrubs; especially for the blossom and fruit of the numerous species of Wild Roses, the blossom-rich Hawthorne and the modest Schiessbeere, at which the attentive nature lover notices with pleasure that it carries beside each other, at the same time, the honey-rich blossom, the still unripe berries, first green, then red, and the already ripened black berries, so that it offers food for insects, birds and the browsing deer as well. And how much fragrance do such shrubs offer! A particular advantage of the beautiful **Sweet Briar** [*Rosa rubiginosa*] is that even its foliage has a nice fragrance.

However, there is an unfavorable connection to wildlife for Hawthorne, European Cranberrybush and *Euonymus,* inasmuch that these woody species unfortunately are frequently eaten and spoiled heavily very often by a moth-caterpillar, which builds webs and cocoons.

Coming down to poor soil classes, once again I have to mention Sweet Briar and Schiessbeere, which like to socialize with **Juniper** [*Juniperus communis*] and **Broom** [*Cytisus scoparius*].

When Shakespeare in *Midsummer Night's Dream* made Theseus speak:

> *"Or in the night, imagining some fear,*
> *how easily is a bush supposed a bear,"*

he certainly thought of the Juniper, for no other shrub is as **variously shaped** as this one, and herein lies its main advantage. The female bushes are particularly beautiful when they abound with blue berries between the green needles.

Growing next to the serious Juniper, the Broom decorates itself in May with the golden splendor of blossoms; in winter, however, "its continuously green branches keeps alive the picture of the vegetation's survival among ice and snow."

The less conspicuous Broom species offer similar services. Where its yellow blossoms are joined by the rose-red switches of the **Restharrow** [*Ononis fruticosa*] at the wood's edge, there are very beautiful color contrasts.

As being particularly able to grow in deep forest shade, I combine the wild ***Ribes*-species** (Red and Black Currants and Gooseberry) [*Ribes silvestre, nigram, and uva-crispa,* respectively]. Like all the shade plants, they become green very early, in order to use the first spring light before the canopy of leaves above them gets dense.

Due to this advantage, they can be appreciated; in addition they are used as a bee-pasture.

The heath family plants, including Heather [*Calluna*] and Heath [*Erica*] in several species, Whortleberry [*Vaccinium uliginosum*], Bilberry [*Vaccinium myrtillus*], European Cranberry [*Vaccinium oxycoccus*], Cowberry [*Vaccinium vitis-idaea*], Bearberry [*Arctostaphylos uva-ursi*], *Andromeda* in two species [*Andromeda glaucophylla*] and Bog Rosemary [*Andromeda polifolia*], and finally Wild Rosemary [*Ledum palustre*], form a sharply distinguished family. I am unable to explain the indescribable charm which lies in these modest plants; this gap has to stay open still, like many other.

Heather plants are the most atmospheric plants; they address more than other ones to the soul. So Prince Pueckler writes in one of his charming letters: "If you like forest shade and solitude and the thousand-voiced singing of innumerous birds, and, when with setting sun the living nature is also sinking into sleep, the mysterious rustling and whispering of the trees, which are bowing and caressing their crowns to each other high above you—then come here, and you will experience blessed moments. Lying on a lush, green, velvet-soft carpet of Blueberry-herbs and Moss, twined around with Wild Rosemary and Fern herbage, I have already dreamed away some sweet hours of my life, until a timid roe passing by reminded me to go home."

Now there are a great number of shrub species, which are not widely distributed in general, but are of importance locally. Here in my area, the very fragrant **Daphne** [*Daphne mezereum*] decorates April with rose-red blossoms, and individual **Barberry shrubs** (it is well known that these refuges of damaging fungi are tolerable only far away from fields) catch the eye by their picturesque branching, especially when they abound with yellow blossoms or blood-red berries.

More important than these is the **English Holly** [*Ilex aquifolium*], which unfortunately is missing in the Silesian forests; this magnificent shrub, which flourishes even in the darkest forest shade, "which represents here with us the laurel form of the South in one of its most beautiful shapes." The holly is particularly beautiful when its scarlet berries shine out between the gleaming, strongly armored leaves.

There are many more shrub species to be mentioned which are locally limited but on their natural site immeasurable; one of the most important is the **Dwarf-pine** [*Pinus pumilio*], whose picturesque bushes decorate ridges and slopes of our Riesengebirge, as well as the different species of **Honeysuckle, Serviceberry** [*Amelanchier ovalis*], **Reinweide, *Acer Monspessulanum*** [Montpellier Maple], **Cotoneaster, Shrub Birch** [*Betula fruticosa*], **Green Alder** [*Alnus viridis*], **Seekreuzdorn** [*Rhamnus frangula*], **Wayfaring Tree** [*Viburnum lantana*], and the **Goldenchain Tree** [*Laburnum anagyroides*]. Unfortunately it seems to be not permissible to plant this wonderfully blossoming shrub beyond its natural distribution, because it is so very poisonous.

Raspberry [***Rubus idaeus***] **and** Blackberry [***Rubus fruticosus***] should be mentioned from the semi-shrubs. Blackberries in all their species deserve special attention. It seems to be more known to painters than to foresters that some blackberry species take on magnificent leaf colors in fall, while we foresters, admiring foreign tree species, often underestimate the native flora. Semi-evergreen, some blackberry species decorate the winter forest, bringing it to life, while offering browse to wildlife.

The German forests are not rich with **vines.** The more one should appreciate the existing ones.

Dr. C. Bolle praises the sites of the Mark, where ***Hedera helix*** [English Ivy] still occurs in abundance, with warm words: "Like shrines of nature, viewed only by few, these forest scenes present themselves, in which the Ivy, having become a tree itself, climbs up another trunk in full freedom."

The most wonderful thing about English Ivy is its double-shaped character, that the upper shoots grow freely into the air as structured branches in old age and carry there very shiny leaves with entire edges. Furthermore, it is strange that, in September and October, these branches cover themselves with blossoms which are externally inconspicuous, but the more attractive to honey-bees, so that the nature lover is made aware of the ivy blossoms already from far away by the loud humming of the industrious insects.

Less magnificent but more cheerful than the Ivy, the fragrant **Climbing Honeysuckle** [*Lonicera periclymemum*] and the **Waldrebe** [*Clematis vitalba*] grow rampant with white blossoms and silver bunches of seeds in the nicely pinnate foliage.

6. Foreign Tree Species

The remarks above have grown disproportionately long, for I have not succeeded in being as brief as I had intended to be.

For my excuse here I refer once again to Dr. Bolle, who talks just as I feel when he says:

"The botanist occupies himself with the plants, because he loves them; but his affection towards them is doubled in certain cases when he sees their foliage and their blossoms come out of the soil which is most dear to him because it is the soil of his home."

While working on the compilation above I became conscious, even more than before, of which treasures are native in our forest, and how much I myself have neglected its care up to now. I think I will hardly be able to find enough space in my forest to give enough opportunities to the native flora, and therefore in the future I will feel even less than before the need to use foreign tree species for enhancement of the forest. Nevertheless, here are some lines that may be dedicated to them.

The **Oaks** take the first place among the numerous tree species of America and Asia which have been recommended for experimental cultivation, inasmuch as they distinguish themselves by the magnificent fall color of their foliage.

But there is a need standing against these advantages: their branching does not show the gnarled forms which give the picturesque appearance to the native oaks.

After once my attention had been drawn to this fact, I found this perception confirmed in every case, so for example near Potsdam and in the Augarten near Kassel. The whole arrangement of foliage corresponds with the branching, and therefore, in winter as well as in summer, the beauty of *Quercus rubra, coccinea,* and *palustris* remains less than the magnificence of *Quercus robur* and *petraea.* Only in fall for some weeks are they dressed more splendidly, but still not more beautifully.

I do not really understand why *Quercus alba* [White Oak] is not used as a decorative tree as often as those other three species. This species, which in summer is hardly distinguishable from our English Oak, colors itself particularly beautifully dark-blood red in fall.

Regarding the **American Walnut Trees** [*Juglans*], I have not yet been able to discover that they are more beautiful than our native Common Ash and its numerous varieties. The **American White Ash** [*Fraxinus americana*] possesses, however, the advantage of beautiful yellow fall color, but very soon after the color change, the leaves fall down and the crowns are standing there bald, while *Fraxinus excelsior* is still long resplendent in the most wonderful green.

Black Locust [*Robinia pseudoacacia*] and **Horse Chestnut** [*Aesculus hippocastanum*] are magnificent trees, excellent in foliage and splendor of blossoms, but they differ too much from the character of our native forest trees. Whoever plants Black Locust should not fail to accompany the original form with the garden variety *Robinia pseudoacacia decaisneane.* This variety grows as fast as the common Locust; its numerous blossoms are characterized by a vivid rose-red color, and they are preferred by honey-bees.

Already Gilpin has noted that the crowns of Black Locust often suffer from breaking. He complains that one cannot rely on it, for the branch one admires today may be ruined already tomorrow.

As far as I know them, **American Linden** [Limetrees or Basswoods, *Tilia americana and heterophylla*] do not arrange their foliage as beautifully as the German Limetrees do; some (the Hungarian Silver Linden, [*Tilia tomentosa*]), however, possess a very beautiful golden yellow fall color of the leaves, and for most of them it must be counted as an advantage that they bloom very late and therefore may be able to help in the extension of the limetree blossoming period. However, its blossoms are not visited as often by honeybees as the blossoms of the small-leafed German limetrees.

I would like to mention **two Alder species,** because they have been widely distributed here for many years. The **White Alder** [*Alnus rhombifolia*] here never attains the beauty that it has in its Nordic home. It deserves some attention only for its early blossoms and its light bark. Much more beautiful is the **Hazel Alder** [*Alnus serrulata* Willd.]. This one is characterized by beautiful dark foliage. It is desired for game food because it will not grow tall, produces very much seed, and stocks itself very densely from root sprouts.

Foreign conifers have still not been tested much in the forest. I know only two of them well enough to dedicate some lines to them.

The **Weymouth Pine** [Eastern White Pine, *Pinus strobus*] is decorated very nicely by its bunches of shiny needles in its youth, therefore it is very accurately called **Silk Pine** by Weise. In old age, the unity of its dark stands gives a magnificent impression. Among all the tree species, this one has the most delicate voice, if I am allowed to express it this way; for it sounds very softly with the slightest murmuring of the wind. It remains behind our common pine inasmuch as its trunk will not dress itself into reddish bark. Furthermore, I have to list to its disadvantages that its top will get disfigured by attacks of the white pine weevil. That is why it should never be planted near places for timber collection and storage.

I planted the **Pitch Pine** [*Pinus rigida*] here in the open areas as a game food because it sprouts back from the cut trunk; I hoped I would be able to keep this game area dense as a low forest. But I could not make up my mind to cut it down, for the trees had developed so picturesquely beautiful. The rich needles keep a nice green even in winter, when our Pine takes on unsightly colors. The large violet catkins and the numerous large cones present a characteristic adornment of this tree species.

Douglas-fir [*Pseudotsuga menziesii*]. I have seen the **Douglas-fir** up to now only in two individuals on Dr. Bolle's Island Scharfenberg in Lake Tegel. This is not sufficient for a judgement. In youth the Douglas-fir is more delicate than the Spruce.

Chapter Five.

Fragrances and Voices of the Forest.

Several chapters have been dedicated primarily to impressions which the eye receives; now I will try, by some short remarks in the following passage, to cover at least somehow **what the forest is offering to the senses of smell and hearing.**

I know well that aestheticians have thought themselves obliged to estimate very low the pleasures mediated by the sense of smell. Their conclusion is this: Smells cannot be defined in words, therefore one cannot write large books about them, and that is why they are not worth anything. However, the ruling taste of educated and uneducated people gives a different judgment. I have watched very calm people becoming morally outraged, when they noticed a lack of fragrance of a rose (the proud Victor Verdier, which looks so good with its large light flower in front of conifers). This appeared to them almost as a betrayal, as a sign that times are going to get worse.

The aesthetician Bratanek covers this extensively, for in his work, which I have already mentioned several times, he dedicates a whole chapter to the fragrances of plants, in which he presumes to claim that there is, given by "the fragrance, so to speak out of the inner heart of the plant, a simpler, faster, sharper recognition in its kind than their shapes and all the attempts of an artificial description." This seems to me to be said rather effusively, but a good deal of truth lies at the bottom of the remark.

The previous remark is true in regard to the forest's fragrance, inasmuch as the smell is different according to place and time, and most characteristic, because tree species, site, temperature, sun and season of the year influence the forest's fragrance in a noticeable way. Winter leaves us starving in this respect, too, for the tannic acid smell of freshly-cut oaks is a doubtful pleasure, and indicates to us the cutting area already from a distance. The more joyfully we then welcome the breath of spring, which, after the first warm breezes, is expired by each clod of earth turned over by the spade and each moss cushion stepped on by our foot. Along with the season then, the fragrance gets richer and richer. First the early Goat Willow, then the small Eared Willows, abundant with blossoms, contribute their best; Hairy Birch and Kienporst are the ones in June, limetrees up to the end of July, and Immortellen in the height of summer, which give the color to the fragrance, as I would like to say. All the time the source of the fragrance is an inconspicuous, hidden one; different fragrances melt into each other; that is why we do not attribute it to the individual blossoms, leaves or resin drops; it is the forest itself, as an individual entity, which seems to flavor the air of life to us.

Quite similar it is with the voices of the forest. Beginning with the first warm beams of the spring's sun, which inspire the finch to its joyful cry:

> *"Break, you stone heart,*
> *For spring will come,"*

up to the time when the powerful bass of the rutting stag sounds, there is something new to listen to nearly every day, and even the silent winter time is vivified by the busy woodpecker.

But the voices of wildlife are—at least during the day—more voices in the forest than of the forest; the forest itself makes itself into the sounding instrument of the wind and the air, as it makes visible to the eye, too, the waves of the sea of air by raising and lowering its crowns.

The roaring of the wind in the tree tops has always been a symbol for the mind, as in the Bible, First Kings 19,12 (Elias) and

John 3, 8. Some verses out of Klopstock's *Spring Celebration*, the ode of the pious poet of the Messiahs, may find their place here:

13. Breezes that blow around me and breathe tender
Cooling upon my glowing face,
You, wonderful airs,
Were sent by the Lord, the Infinite!

14. But now they become silent, they hardly breathe,
The morning sun gets muggy,
Clouds stream up,
Visible is, who comes, the Eternal One!

15. Now they float, they roar, the winds whirl,
How is the forest bending, the stream rising!
As visible as You can be to mortal ones,
Yes, this is you, visible, Infinite One!

27. Look, now Jehovah does not come any more in the weather,
In calm, soft murmuring
Jehovah is coming.
And below him the bow of peace is bending!

Now I would wish that instead of me, a professional colleague with knowledge of sounds is going to describe the voices of the forest, and I think such a description might be not totally unfruitful. How many column-long discussions of everyday concerts do the newspapers print, certainly with the success to educate the reading and listening audience to better understanding and increased pleasure? Should the same striving towards the concerts of nature remain in vain? Only practice, totally without instruction from outside, was sufficient to bring a person as unmusical as me to listening to the voices of the forest with pleasure. However, I have not arrived here completely without training, but this did not come to me out of books or from teachers, but by the possession of one of those aeolian harps that are produced by Carl Jordan in Potsdam as instruments rich of strings and of charming melodious sound. Since I practiced with the harp to listen to the tender up-swelling and dying of a sound, the coming and going of the most different sounds, some pleasures, which escaped me otherwise, reach me in the forest, too. Now I do not stay indifferent anymore when my Quaking Aspens, individually scattered in a young beech pole forest, soon raise their voices from the distance with the first nearing of a breeze, when they accompany it and then grow silent, as soon as it is over, when they then make their whispering noticeable in the entire stand in the regular blowing of the wind, until they are overpowered by the roaring and rustling of the young beeches, which bow to stronger air waves.

On the foundation of such observations I would like to attribute to the wind in the tree-tops three voices of not only different intensities, but remarkably different character:

In the soft blowing of the wind the leaves of long-stem tree-species pat against each other (**the whispering**), then the rubbings of soft branches are heard (**the rustling**), stronger breeze sets every leaf and every branch directly into sounding vibrations (**the roaring**). I thought I was on the right track with this division, then afterwards I discovered that Schleiden divides quite similarly, only he puts even before the "strange whispering, under which one tries to lay words spontaneously" the "quiet, uninterpretable murmuring." But I think that one cannot hear the "murmuring" with less breeze than the "whispering," but it demands some more wind. One can only hear it with conifers.

I remember reading in Moltke's *Letters From Russia* that there are bands there, in which each virtuoso knows only how to play one note, but this one quite masterfully. Similarly, nature has trained its orchestra. The most delicate voice of all was given to the Weymouth Pine, the actual roaring we can hear in the most magnificent way in the pine forest, and the whispering sound is especially characteristic of poplar species.

It is an advantage of mixed stands that one can hear all three voices sounding at the same time, or following each other in quick succession.

Final Remarks for PART ONE

The first part of this book, in which it has been taught that the application of forest aesthetics is an artistic performance, is apt to discourage the reader. For an artistic performance, special personal talent is necessary, and this does not need to be proved; genius is necessary, and this is a gift of God, which is given only to few favored mortals.

Whoever in other fields is going to work without brilliant talent, only according to rules, will, bound by them, produce only mediocre things, for his work will remain unfree; but the forester does not need to lose heart because of that. What he lacks is going to be replaced by our friend, nature, which never allows itself to be forced completely under the yoke of rules and which rewards our love with answering love.

Even if we do not act like a genius, but simply, reasonably, and understandingly according to studied rules, the freedom of nature will heal the mistakes we make. It is standing at our side as a teacher, as an assistant, supporting and improving.

So nobody should, out of timidity, refrain from caring for beauty in addition to the material benefits of the forest.

The second part of this book is supposed to teach how to do that.

PART TWO

Applied Forest Aesthetics

SECTION A. Forest Establishment and Forest Economics.

Chapter One.

The Determination of the Most Useful Type of Soil Use.

The most important question, which the Agencies for Forest Organization [Forest Establishment Organizations] as well as the civil servants entrusted with the work within the forest districts have to answer, is the question for the most appropriate type of soil use; namely whether the **existing forest area** is to be maintained in unchanged size, whether it is to be **enlarged, reduced** or **distributed in another way.** Having been suggested from time immemorial and answered in very different ways, this question stands again in the center of current interest. The social politicians have already frequently dealt with it. One may hear how vividly E. M. Arndt in his time had already taken a stand in this matter. He wrote:

"Foresters and other brave people have written for the wild animals of the forest and the cattle of the stables, for deer, roe and hogs and for horses and donkeys and oxen; they have written also on forests, pastures and fenced enclosures—how to lay them out and how to manage them so that the wildlife and the cattle within do well and thrive, and that their noble blood and their good character are not going to be spoiled. Once I have thought about the single one left alone, poor human, how for him the earth would need to be fenced, forested and managed. And I have not thought, but felt it, and, wherever I stepped it has—I am not using a vulgar expression—dropped upon my nose and has jumped into my eyes . . . therefore they have to stay and they need to be recreated, the old German groves, **trees do not need to lack the German human anywhere,** with whose branches he can plaite his arms like with as many arms and with whom he can merrily swing himself up to his stars."

Then his suggestion is to **plant with trees all the mountains,** "as if they were sacred like the old groves of the gods" (the valleys may stay dedicated to cropland cultivation and pasture), the German **plain, however, to be run through by forested strips,** which have to be at a maximum distance of 1.5 miles from each other, being at least 1,500 feet wide, and which should never be clearcut. Arndt thinks this will result in not only an improved climate, but also more ideal profit; for regions "which lack the proud green and adornment of the trees" appear to him only to have been made "in order that the poor great-grandson of Adam can eat his bread with sweat upon his face, and creep along among thistles and thorns, his eyes only directed downward to the furrow with his bull and not drawn upward by a more cheerful and more free attraction of nature in his eyes and heart."

Later Riehl discussed the same question in a really exemplary way. "Even if we did not need any timber anymore," he writes, "we would still need the forest. The German people need the forest, like the human needs wine, although it might be completely sufficient for the pure necessities of life, if only the pharmacist put a quarter of wine in his cellar. If we do not need the dry wood any more to keep our outer human warm, then we will need even more the thriving lush green wood for warming of our inner human."

In order to adequately meet what Riehl demands of the woods, the forest has to extend over a large connected area, it needs to be managed in a big way, for only such a forest (I again quote Riehl's words) makes **"us civilized people still enjoy the dream of**

personal freedom, untouched by police supervision. At least there one can still walk crosscountry according to one's own wishes without being restricted to the beaten track. There a grown-up man can still run, jump and climb after his heart's wishes, without the precocious Aunt Decenz thinking of him as a fool. Fortunately these relicts of Germanic forest freedom have been saved nearly all over Germany. In some politically freer neighboring countries, where the fatal fences finish soon the unbound lust for hiking, they are not known any more . . . How do their liberal laws help the Englishmen, when they have only fenced parks, when they hardly have a free forest left? The force of custom is unbearable to a German man in England and North America . . . Poetry has called the free forest and the free sea in a profound word also the sacred forest and the sacred sea, and therefore nowhere is this sacredness of unspoiled nature more touching than where the forest emerges directly from the sea. Where the pounding of the breaking sea roars together with the rustling crowns of the trees into one hymn, but also in the soundless noon silence of the German mountain forest, where the hiker only hears the beating of his own heart in the wilderness, being miles away from any human settlement, there is the real sacred forest."

Schleiden, too, has established the need for **large, connected forests.** "There are two forms of nature, which, although seemingly so different from each other, have as an effect moods which are related to each other; these are high mountains and vast virgin forests. As the view in the high mountains penetrates the infinite distance, in the virgin forest it is limited to the nearest things, without, however, being impeded by a definite end, for example by a wall of rocks; so the common situation that causes the general atmosphere in both is the isolation of the human, and the feeling that he, all by himself, faces the entire nature with its eternal forces, its eternal quiet working, that he feels small and independent from the great but nevertheless also big as a living part of the whole, that he can abandon himself to the sublime law of nature in an elevating security of confidence, always and unchangeably working in the same way, safe from any disturbance and confusion, where intelligence, which he had to use in contact with humans in order to maintain his existence, is as unnecessary as it is powerless. . . . The forest ends where the actual social human life begins."

Now, how large does the connected area need to be in order to serve these purposes? I would like to answer: if it is large enough for deer, then it will also be large enough for the human who is looking for **solitude** in the forest.

The closer humans are together, the fewer trees they need, but Riehl wants to grant a forest to each village. To him "a village without a forest is like a city without historic architecture, without monuments, without art collections, without theater and concerts, and above all, without comfortable and aesthetic inspiration. The forest is the gymnasium of the youth, and often the festival hall of the old people. Is this not at least as valid as the economic question of timber?"

At the moment, it is not only the voices of individual, especially active minds, which demand the creation of vast new forests, as Arndt and Riehl did years ago; **nowadays the desire for extensive afforestation is rooted deep in the people's consciousness.** Numerous proclamations in the newspapers and in the parliaments prove this; the administrative agencies, too, are open to this task. With foresters, too, their desire for afforestation is not exclusively rooted in the wish to grow more timber or to improve the hydrological situation; they are also founded on more ideal attitudes. "The afforestation of wastelands" (I quote Kessler), "is so to say the atonement for mistakes already made in the past, looked at from a national-psychological point of view, and the vivid desire for this action is, however subconscious, not without a certain regret over ruthless intervention on conditions created by nature, not without a certain worry about the bad consequences of these disturbances. It is to be put right what the shortsighted economy, namely one-sided egoism, has caused. . . . The greatest, however more psychological, influence on the striving and demand for afforestation has, among others, a certain more aesthetic than economic **sense of order,** which is particularly characteristic to the German man, who is used to regular and orderly conditions. When an order-loving eye sees vast uncultivated areas, or areas grown only with juniper bushes and birches, so it is understandable under this point of view to feel the need for and to express the wish to see regular plantations and forest stands instead of this picture of waste heathland, which, more than anything else, shows after all that here the ordering hand of the human has intervened and was effective."

The quotations above show without doubt, that the **question of afforestation has to be decided quite particularly after considering aesthetic points of view and is actually decided accordingly,** as the last sentences of Kessler, quoted from his afforestation practice (in the government district of Danzig) really show. Sprengel as well reports from the district of Rosenthal in Holland that people there are working with eagerness on the "afforestation especially of heathlands, less in order to achieve high revenues, but in order to get rid of the sad picture of the country."

I quote the words of others because one might be inclined to doubt the voice of the professional forest aesthetician as being not impartial enough, although I am going to be surprisingly modest in my demands. The percentage of forest area that I regard as the necessary required minimum is a low one, because what matters to me is **a more suitable distribution rather than more quantity. I wish the forests were distributed in such a way that one could see at least one from any place, be it only at the horizon, and that it might be possible even to the hiker to make a forest trip from any place, leaving and returning on one day. For this, one single forest is necessary within about each five square miles, and if we gave it a size of one quarter square mile, this would make up only about 5% forestation.** In addition, Germany has to still keep its forest areas. Therefore the large forest complexes, existing at the moment, especially the historic old bannwaelder [protected forests] should mainly be maintained, so to say (using Riehl's term) **as sacred forests;** for, in order to make the human able to feel being alone with nature and God in the forest, these 6,000 acres in one unit, demanded by me above, are not nearly enough. This modest demand is valid only for the plain.

I am leaving the **mountain sites** completely out of consideration, whether **all the mountain forests** serve very important purposes as barrier forests, or not. Even if I myself subscribe to the first attitude, I am going to consider only the **purely aesthetic point of view.** From this point of view, I am certainly not going to suggest to bury **every** fresh flowering alpine pasture under a dense spruce plantation, but I think **I have to demand tree growth for every mountain sufficient to give a forested impression from the distance.** To the impartial mind, the hills and mountains appear as the **natural shielders** of their region. They lose this charm as soon as they themselves are lying bare, unsheltered from all the negative effects of the climate. The mountains are to us an **attraction,** a goal of longing. They lose this charm, also, when they, instead of letting a meadow shine through only every now and then, or a cliff, offer already the distant viewer everything that they have to show. The demands of the **theory of color** are important here also. Shaded by the perspective, forested mountains melt with the vault of heaven, and on the other hand, the view from the mountain down into the valley creates a picture, so to say, only if there are some framing tree tops in the foreground.

Up to now I have not yet seen a true **heathland landscape,** and so I cannot discuss the aesthetic necessity to plant these with trees. However, one can grow fond of the heathland, at least the Lueneburg Heathland, without having seen it; for that, one only needs to read the appealing description of this region in Burckhardt's Out of the Forest V., written by Forest Master A. Meier; and therefore it surely would be regrettable, if the **whole** Lueneburg Heathland itself would cease to exist.

We need distributed everywhere **smaller forests,** besides the greater and medium-sized forest areas and connecting elements, so that (as E. M. Arndt demands) "nowhere trees lack to the German human."

Strangely enough, our contemporaries (the same ones which express such a vivid interest in the **great** and **difficult** task of blessing regions that are poor of forests with the establishment of larger connected forests) regard the treasure of small copses and shrubs with indifference, as if it had to be like this, that they disappear more and more from day to day, without even the attempt of protest. This indifference is the more strange, because in addition to the aforementioned social politicians, poets (Goethe), physicians (Dr. Zwierlein), foresters (von der Borch), garden designers (Prince Pueckler), each one in his way, have pointed out how small copses, adequately distributed and suitably joined, give aesthetic profit in addition to valuable protection to the environment of every house.

In his time Cotta already had regarded it as being a special advantage of his **tree field cultivation** [combination system of forest and farm crops], that it would spread the tree cultivation into areas without trees prior to that time and would brighten these up. He has no hesitation to quote an "appreciable essay" out of the *Economic News* of the year 1811, which concludes with the words: "What an idea, what a look, if within a few years the whole monarchy would be transformed into an earthly paradise! Everywhere pleasure and purpose! Everywhere shade, shelter and harvest! Timber against freezing, fruit for eating and refreshment, sugar for the taste (he means maple sugar! the author), spirits of wine for strengthening—all the voyages in the milder seasons just walks through an immeasurable garden!"

A nice picture of imagination indeed! Cotta has flattered himself with the hope that this "picture will become reality" by his tree-fields.

If it is now clear that the realization of Arndt's suggestion regarding the wide forest strips would not have brought such great advantages that they might have been able to be regarded as a substitute for the necessary sacrifices, then I think that the straight-lined tree rows of Cotta could not have contributed particularly to the adornment of a landscape at all.

What has been good in the suggestions of E. M. Arndt and Cotta has already been achieved at many places, partly by the **knicks** [hedgerows], which break the power of the storms in Schleswig-Holstein, and at other places, partly by **"shelter plantings"** [shelterbelts]. These have been established extensively, particularly in the high Westerwald. These, too, serve the purpose of wind protection for humans, cattle and fields. At the first Main Meeting of the German Forest Society, the following was announced concerning this matter: "Around 1840, the shelterbelts have been planned extensively by the commendable sponsor of Westerwald cultivation, Privy Senior Civil Servant Albrecht (at that time in Emmerichenhain), on all the mountain ridges and saddles not yet forested and along the roads and the property lines of the individual community pastures in order to interrupt the air flow; being narrow forest strips 10-20 m wide, containing spruce, alder, ash or beech, they were actually carried out at that time to a considerable extent in spite of great resistance from the population. In 1869, a "General Cultivation Plan" for 64 parishes of the Westerwald was drawn up, paid for by the state, in which for each one of the parishes the most suitable position of the cultivation types (as a rule meadows in the hollows, fields and pastures along the slopes, forest on the crests), of the shelterbelts and the drainage and main road system according to standardized points of view were suggested as a basis for consolidation. In the forest report of 1872, given on the finished cultivation plan, it is stated that according to the experiences up to that time, only spruce *(Picea excelsa)* could be considered for afforestation. . . . In addition to the advantage which the shelterbelts offer in these rough mountain ranges of a welcome protection against the wind, there is the value of their direct wood revenues, which leads to their continuing recommendation. The spruce, however, does not earn the particularly high profits of the dense pure spruce stands in the Westerwald, only because of the limbiness of the many spruces on the edges, but they force from the poor soil nevertheless an income which it does not give for any other type of use.

The rotation of these shelterbelts has been set at 60 years, according to their main purpose, for the narrow wood strips allow more and more wind with growing age; but the cutting always has to be done only in half the width if they have a width more than 10 m-if they have a width of more than 30 m, only in 1/3 of the width-and always with the saving of the deep-rooted and low-branched edge trees. Shelterbelts of the same age and more than 10 m wide are to be partially cut before reaching the rotation age, in order to arrange gradually the normal sequence of age classes in strips at least 6-10 m wide and with a maximum of 30 years age difference.

According to the special records maintained for the shelterbelts, (that means only the **narrow** wooded strips of the high Westerwald, established as a wind protection, mostly running as long ribbons from one forest site to the other one) they now contain

11 ha in widths of less than 10 m
81 ha in widths of 10 to 30 m
360 ha in widths of more than 30 m

---

for a total of 452 ha.

The **hedgerows** have the aesthetic fault that they offer few changes and they make the whole region broken. Up to now I have had no opportunity to see the **shelterbelts.** Anyway they are going to make a good impression on mountain slopes, when they run according to the contour line, but otherwise they would give quite an unnatural striping to the mountain slopes, which would shape the character of the landscape disturbingly.

From the aesthetic point of view, the **Pueckler hedges** are ideal. In his time [c.1820], Prince Pueckler recommended as an example the **planting of the road edges with irregular tree and shrub structures,** which he had seen in England, and this hint has often been used. Such hedges present an essential part of **open areas,** which will be discussed in detail in the third chapter of Part II B.

It is always easier to save existing things than to obtain missing parts. **When transforming a forest into field or meadow,** therefore, one should see from the beginning that suitable places (there will always be some, which will only be able to serve another purpose after particularly expensive ameliorations) are going to be maintained for timber growing.

I am reminded of some land compensations for the purpose of Servitutabloesung [release of obligation] as an example of how it ought not to be done, where the forest administration has even neglected to secure for itself a **shaded road** (pollarded willows offer very little shade) across the farmer's land up to the forest.

While the invasion of the forest into the landscape is mostly an advantage for this one, the opposite is mostly evil. **Field enclosures** in the forest always look strange! One would need to feel just an urgent desire for change, being tired after several days of forest hiking, in order to welcome willingly only one large forester's private field, not to mention a whole colony. If possible, one should tend to plant areas like those with trees. But it is a little bit different with the **meadows.** The grass area, every year sprouting freshly green without any assistance from us, not furrowed by the plow, only during the harvest a place of busy action, fits well to the forest's character, with which mother nature, too, does her best. On the other hand, it offers a pleasing contrast by its abundance of light in comparison to the shady center of the forest, as well as by its openness and its walkability; the contrast will be more conspicuous the more welcome it is—the larger, the more closed and less open the forest is. Small forests, especially those whose so-called cultivation fields, being many-years-old wide clearings, usually bring up more grass than we would like, may rather lack the meadows. In pine districts, for example, if alder swamps are situated between sand dunes, one should not be tempted too quickly to cut immediately any alder stand which may not be growing perfectly well.

III.
Nesigoder Thiergarten. Insel mit Erlen.

For **water surfaces,** similar things are true as for the meadows. The contrast that they make with the forest is also a **harmonious one.** (Having both strong contrasts and a strong connection to unity, this relationship is what we learned as the source of richer beauty.) The lightness, the openness is theirs even more than of the meadows; the contrast between the horizontal lines of the water surface and the unifying vertical lines of the vegetation is added as another contrast, but the unifying elements are increased, too. Together with the forest, the water surface rests at quiet times, and together with the forest it moves in the wind, in the storm. It contains rich wildlife as well as the forest, and it keeps the charm of the mystery in its depths as well does as the thicket. **Therefore water surfaces in the forest district—small ones and large ones—are also a real treasure in the aesthetic sense, whose maintenance and additions have to be looked after deliberately.**

Living in the middle of the great pond system of the Militsch-Trachenberg District, I have, almost daily, the opportunity to watch how much artificial pond constructions enhance the area. Now I see the ponds shiningly reflect the sunlight, now I admire the marvelous inks, when the evening sun, going down above the ponds, shines through the humid air; one moment the seagulls come flying over individually or in swarms from the ponds up to the fields, even up to the plantations. In spring and fall the waterfowl migrate above my forest to their breeding grounds and back. The wild geese fly particularly conspicuously, announcing themselves from the distance in a wedge-shaped arrangement, up to the old breeding-grounds, which they inhabit in the artificially dammed "Luge" of the Nesigoder Thiergarten [Zoo].

The photograph No. III shows one of these islands, which the geese prefer to use for their breeding. All this is offered to me by the ponds, although I live 10 and 25 km away from them on a hill. Whoever owns the ponds can enjoy all this even better, and in addition to fishing, waterfowl hunting, and roe hunting along the shore, rides on the dams trimmed with oaks and many other pleasures, besides a considerable economical advantage.

Such impressions make me wish that everywhere where the opportunity exists serious consideration will be given on whether the construction of ponds should be preferred to the drainage of wet sites.

Lake surfaces are even more beautiful than ponds; they usually offer a clear surface all year long because they are mostly deeper and therefore not covered with reeds. That is why it will always be a mistake in respect to aesthetics to drain lakes or to lower their water level.

Chapter Two.

The Design of the Road System, Establishing and Naming of the Administrative Units

The layout of the road system and the establishment of administrative units are works which already have to go hand in hand with the decisions discussed **up to now,** as they present the frame for the allocation of the existing or proposed forest stands into the appropriate management classes. Directing the procedure according to older textbooks, one would have to first decide **whether to choose the regular or the natural division.** However, to us the names of the different methods are already bound to provoke aesthetic reservations, for should something be desirable which is not natural?

During the last decades, the natural division has been recommended so effectively in several excellent writings that I had intended to omit from the second edition of *Forest Aesthetics* the argument against the regular division as now being superfluous. In the meantime, however, Neumeister's *Forest Structures of the Future* has been published, in which this sentence can be found: "The **rectangle,** with the long side twice the size of the short side, is the ideal form, but it can only be implemented in the plain. Sides of 600 m and 300 m are particularly recommended here." I think I cannot let such a well-known defender of the rectangular division stay unnoticed.

Even I admit that the rectangular division of a forest district in level land offers advantages to the land surveyor, but I think of the disadvantages as being predominant, and not only in respect to aesthetics. For ease of management, having all compartments the same shape never offers a practical advantage, and having them all the **same size** offers an advantage only for silvicultural systems in which one puts emphasis, like sometimes with the low forest, on the fact that every administrative unit contains an equal number of equal annual cuts. It is not understood, however, how the same size should help with the high forest. If the compartments contained only one forest soil type, and if they were expected to contain at the same time only one single tree species, all of equal age and equal quality, then one could rely on the idea that it might be a little relief to the future head forester in this future ideal forest that he could memorize the size of his management units, and at the same time the divisions, etc, with only one single number, and then will never again need to look it up in the survey index. This doubtful advantage, however, discounted out of a distant and uncertain future up to the present, could present only an infinitely small present value, and this infinitely small benefit is offset faced by a larger disadvantage. The endeavor to cut as many quite regular compartments out of the body of the forest district as possible leads, however, very often (one can see that on many forest maps) to the fact that the unavoidable edge-compartments receive impractical relations of shape and size.

It is no better with the preference for **perpendicular road intersections.** These are said, and justifiably so, to make the measurement of the cuttings easier, but this is only something very minor, for one will need only a few minutes more to measure up other forms, if their boundaries are all straight lines. If it is said, furthermore, that rectangles develop the forest with the least sacrifice of forest area, then this appears to me to be more wrong than right, right only in the respect, as one thinks only of moving the timber up to the firebreaks; it is the other way around with the further transport of the forest products out of the forest and with the ways of the administration and protection personnel. Then one often is forced to make inconvenient detours by squares and parallelograms as well, if diagonal roads do not allow a short cut. These will be necessary quite often in **two different** directions, and then they are going to waste much more space than had been saved by the right angles.

So, the **equal size** and the right **angles** of the compartments do not offer noteworthy advantages, and therefore there is no reason left to let the firebreaks run parallel to each other, rather than putting them fanlike or starlike, **in case that serves any purpose.**

The last six words are directed against the old French **Carrefour** system. This is too much into the opposite direction, as it deliberately prefers the starlike meeting of many roads in one point and avoids parallel lines on principle, even if no practical interest gives reason for the deviation; while I only speak up for the idea **that one might not want to scorn starlike and fanlike arrangements in favor of the endless monotony of the homogenous perpendicular angles, where, for any reason, they present themselves as recommendable in a special case.** With cautious observance of such principles, the necessary change in the intersections of the firebreaks will appear correct by itself without sacrifice of area.

The clarity of the forest district does not need to be decreased in this, for while the division on the map is clearer the more regularly it has been done, in the forest the endless monotony of homogenous intersections has its disadvantages, too, in this respect. However, there is nothing more convenient than to step close to the compartment corner stone, by which the knowledgeable person can tell at a glance where he is standing now, where the neighbor stone can be found and who knows what else!

But what if then

"the nice numbers unfortunately
have been washed away by the rainy wind,"

or at night, when it can be rather fatal if one intersection always looks exactly like the other one, the memory does not find any clues, and the map does not offer comparable elements.

"Yes, but how about the wind" I ask myself, and it is obvious that, if the firebreaks do not run parallel, only part of them can have the position towards the wind direction which we have previously regarded as being the best one.

To this I respond: that the direction of the main firebreak parallel to the wind direction is very good has been proven by the experience of half a century, but that it is even better if set diagonally to the wind is proved by Denzin in the most irrefutable way. Whichever direction one is going to give the firebreak, one will have either a very good one or an even better one.

From the last paragraphs it is noticeable that I only fight the regularity of the division, but that I might accept straight-lined forest firebreaks under some circumstances. For there is no doubt that in the plain, **firebreaks** run straight for long distances are the most beautiful because they are the most natural; it would be narrow-minded if one would want to break a miles-long firebreak in order to perhaps avoid a small hill or save a bridge; and there is one thing I would like to praise particularly for the straight-lined firebreak: When they cut perpendicularly soft waves of the relief, they provide a benefit which cannot be achieved by curved roads. For we are tempted into an optical illusion by the ease with which the eye perceives the seeming convergence of the parallel road edges. As when fog causes objects to appear larger, so when we stand on a hill, the valley looks deeper and the other hills look higher than they really are; and the landscape gains more "movement" (this is the technical term out of garden design) for us by this than is really inherent in it. Standing in the valley, however, the small piece of sky which shines out under the tree branches above the next hill before us, gives an always attractive end of our field of view, like the landscape gardener aims at, often in vain, with an obelisk or some other expensive luxury at the end of an allee. Such an ending of the forest firebreak with the sky shining through is shown in Picture IV.

IV.
Postel. Danckelmann-Linie.

Whoever takes a good look at this picture will notice that the Danckelmann Line is wider in the foreground than in the end. The photograph shows this, but in reality even the sharpest eye may be deceived. One is tempted to assume that the forest is going to continue infinitely, while the part of the line visible in the picture is only 280 m long. The firebreak, however, is cut open 3.60 m in the foreground, but 3.00 m in the background. The intended illusion is aided by the fact that the trunks in the background have stayed smaller and thinner as the elevation of the land increased.

In some cases there is an opportunity, too, to achieve even better results by opening up **vistas to special points,** like castles, towers, etc., by cleverly led lines. Already Grebe, in his management regulation, has recommended using such conditions. But in a small forest one had better avoid the vistas in order not to reveal too early where the forest ends, and one should consider how the matter looks from the outside. A main attraction of the forest lies in its mystery and therefore it is not advantageous for a straight firebreak to offer a view into the center of the forest from its edge. For this reason, one should not lay out a forest firebreak as a direct continuation of a straight road, but rather break the line a little bit at the wood's edge. If, however, a road is running along the wood's edge and the forest firebreaks are coming in from the side like tributaries, then out of this there will be no disadvantage, for the interest is increased particularly by the brief glance in passing-by.

If the foregoing discussion has shown that I am not an enemy of the straight line, I nevertheless think that even in the plain the preference for it has been applied too often. Quite a few old roads have been straightened with relatively high sacrifices, although their moderately curved path might have provided a pleasant change without a noticeable detour. These mistakes belong to these untimely outbursts of a desire for enhancement, which the people cannot avoid, particularly those who are addicted to the utmost prosaicness of attitudes; for only rarely will the wish to shorten the road's length and to facilitate the surveying have been the real reason for these errors; rather the rational, practical man recognized the straight line as being generally practicable and therefore regarded it as being beautiful, and in the same way he thought of every curved road as being everywhere ugly and therefore straightened it out, even if in certain special cases disproportionate sacrifices resulted.

By the way, inasmuch as **small problems of relief** which one tends to by-pass in the division of the districts mostly by curved roads, which may give reason for leaving the straight line also in the regular division, will always be able to be decided only according to the special situation. I am more inclined to try to give some clues, as I cannot agree completely with the principles on this matter as given in Muelhausen's *Road System of the Gahrenberg Forest District.* For there it is stipulated that where division lines would be too steep, one could replace them by curved lines, rising at 6%, "which of course have to be located in a way that the curve is as small as possible and the straight direction will be regained as soon as possible." I can respect this principle as being right only with reservation, namely that the deviations from the straight line should never be so small that one might think the deviation from the otherwise dominant straight line happened only out of inattentiveness. One might try better to remove the bad place by cut or fill, or, as much as necessary, lead a part of the road around the hindrance, while the firebreak itself, as a borderline of the management unit, follows the straight line without interruption. Where the first solution would be too expensive, or the second one would result in too much loss of area, then one might better choose a **considerable** deviation from the straight line in the beginning. Several times I have used a little trick: by breaking the line on its highest point. The figure [22] will clarify this method.

Considering the illustrated case, one would want to establish straight-lined division from A to B, but the hill C, the highest part of a row of dunes EF, would present too inconvenient gradients. One could establish an auxiliary road in the direction suggested by the row of crosses; but it would be shortsighted as well as unsightly if the running of the fire- break would be interrupted, even if it is long, just for such a small obstacle. It would be better to choose the direction AG-GB, because one can not overview the direction beyond G either from the point A or the point B and therefore the mistake will remain unnoticed; this would not be the case with the direction over g, and that route, therefore, is not tolerable.

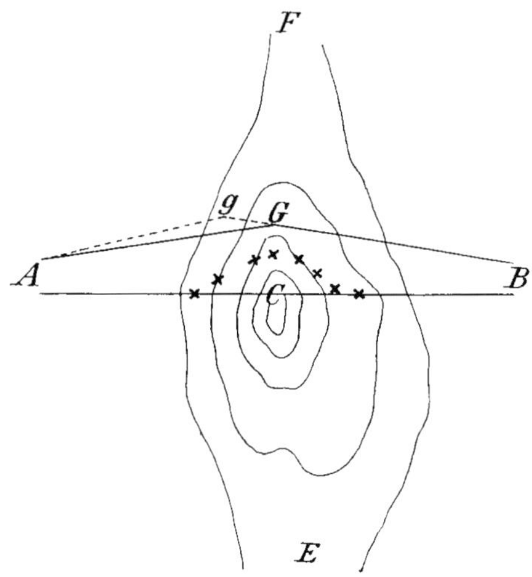

Fig. 22

Quite different is the situation with the direction EF. This line, which runs continuously along the unevenness of the terrain, would give suitable opportunities to the development of nicely stretched curves, and for this situation, the straight line would have to be excluded totally.

I would like to explain the things I said up to now by some examples **out of reality,** however *exempla* (namely those which show how not to do it; they are the most instructive) *sunt odiosa* [examples are annoying], therefore I prefer to construct myself an ideal example, so that no one can accuse me of partiality, that the conditions for the so-called artificial division are supposed to be very favorable. At the rural road from A to B may lie a 200 ha field stretching on both sides of the road from C to D. The area is to be planted with pines and is to be managed on a 70-80 year rotation; that is why it seems to be adequate to create 18 to 20 compartments. The area is level, and no other transport directions than to A, B, C, or D are to be considered. Fig. 23 shows the existing situation.

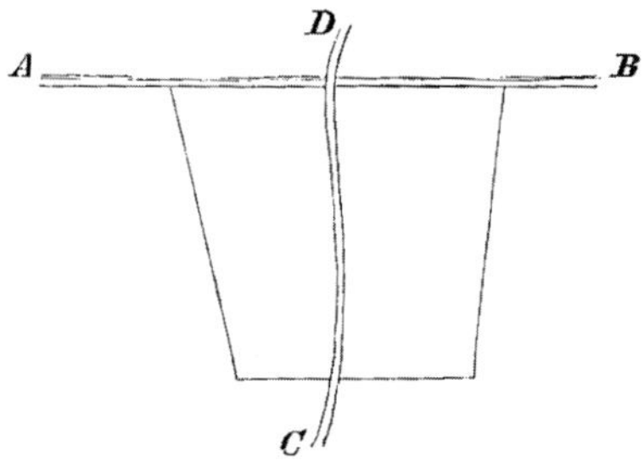

Fig. 23

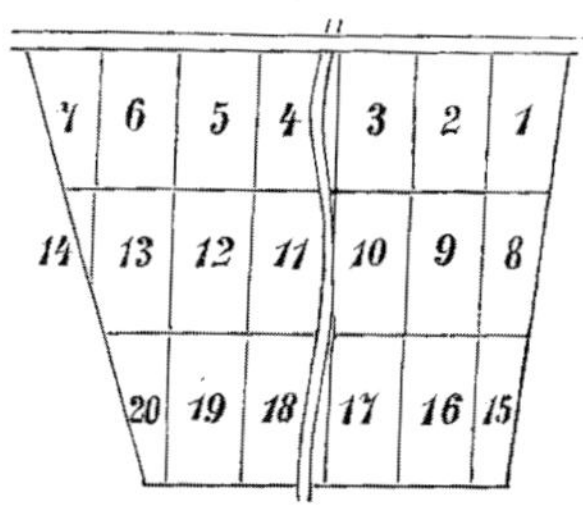

Fig. 24

According to the traditional way, the division into compartments might be done as in Fig. 24. First of all, one would waste some money for the straightening of the nicely curved road CD, whereupon the public driving on the newly opened firebreak will suffer distress (until the soil has settled down), and the forester is going to be desperate, too, because really nothing is going to grow on the previous road area; but a wonderful foundation for a division into compartments has been gained, and we even get—such a gain!—13 of the 20 compartments will be of exactly the same size; it is such a pity that numbers 14 and 15 are going to be so far behind the desired size. On the map now it is easy to work with the angle, and what harm can it do if the timber transports are going along the sides instead of in the shortest direction along the hypotenuse—everything is orderly and clear; what does it matter there if it gets appropriately boring.

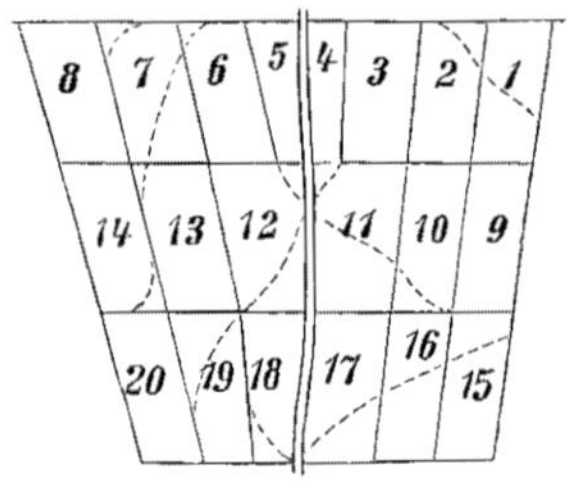

Fig. 25

After this somewhat negative critique of the currently accepted method, I should be cautious about coming forward with a concept of my own; nevertheless, I will try to show how I would do it, in Fig. 25.

In this way, however, more meters of firebreak are necessary than in Fig. 24, and there are diagonal ways, too. But due to that, the forest is quite considerably better developed than a rectangular network might be able to do.

Under conditions which are not as easy as the ones shown here, an inflexible rectangular net often causes heavy economic sacrifices because it can not adjust to the differences of site and the forest stands; it still remains visible in some forest districts even a half century after the establishment of the compartments. With my more flexible way of straight-lined division, these sacrifices can be greatly reduced.

The wide **strips of the Saxonian establishment procedure** are sometimes visible into the distance, and this is unpleasant, especially in hilly regions. Already Gilpin noticed accurately: "All divisions offend the picturesque eye heavily; it prefers to wander around freely. Furthermore, the beauty intrinsic to a forest is increased very much if the great lines of nature and the multi-shaped swellings of the ground remain uncut by these invading divisions and are allowed to have their entire play of undulating waves." For the sake of the same consideration, small geometric cuttings will be able to be applied only in an urgent emergency.

In respect to the **natural division** (now I use the term in the more narrow meaning that has been established anyway), I only need to stress that if it is really natural, then it is certainly going to be beautiful. One should avoid only schematism, and not fall into the extreme to reject the straight line even where no circumstances at all call for a curve, and one should not too slavishly follow the rule that each trail has to follow the borderline of an administrative unit. Especially nobody should decide without the most convincing reason to remove already **existing roads which pass through an administrative unit,** for these especially serve very well the most important aesthetic reasons. Walking on them through a stand that is the same on both sides, we feel much more like we are in the center of the forest than when on a road that leads only along the edge of the stand. Therefore it is mostly a mistake to reject an old compartment boundary and accept the unfavorable setting and bad drivability of the firebreak, which will always require sacrifices like every transformation does. It is much better aesthetically and practically to make the compartments usefully accessible by roads leading into their interiors. By the way: the objection which has been raised recently against these roads, that they serve the block cutting only **once,** while the roads along the edges first serve the right-hand clearing, then the left-hand, is not sound, for the first ones are used from both sides at the same time, and 2 x 1 is = 1 x 2. Nevertheless, these "schriem-ways" (like we Silesians say) are being removed eagerly at many places. I remember numerous cases from Royal and private forest districts, where an unfortunate road which went directly towards the boundary through a compartment, was stretched in a merciless way around the compartment. The forest personnel wasted more expensive balled plants, thorn trees of every kind and curses in abundance just in order to harvest nothing at all from all these efforts other than some poles from thinning (for in the early period of regeneration, rarely will something be growing in the hardened soil), except for additional difficulties in timber transport and obvious animosities from the victimized public.

However, the forest organization is going to serve very well the civil servants as well as the public when it prescribes the establishment of special **short-cut paths for hikers,** (access trails, hunting paths, or whatever one might want to call them; they also might be usable for a carriage road of the 4th order, if this can be done without too much trouble) or (even better), leaves it up to the Head Forester's discretion, because these trails and paths can and are to adjust exactly to the needs changing with place and time.

There has been much argument about **whether longer road sections must be laid out with uniform gradients or not.** According to the motto—*variatio delectat* [variety delights]—I would **prefer not to distribute the gradient uniformly.** In this way, we also gain more freedom to lead the road closely to attractions (rocks, beautiful trees, etc.).

I owe a rule, which is noteworthy for mountainous regions, to von Guttenberg. "Where we have the choice with mountain roads to reach a strived-for peak either by switchbacks or by a wider development of the marked-out route over individual valley bays, mountain saddles or the like, the latter solution should always be preferred, also under the aesthetic point of view, because in this way, many different forest sights and vistas are going to be developed. On the other hand, the switchbacks cut through only **one** stand and are at the same time always questionable for the maintenance of complete forest stocking in the often narrow intermediate strips."

Textbook formulas are supposed to help in determining the proper angle with which the tributary roads make in intersecting the main roads on mountain slopes. For this I only have to comment that **too sharp angles are not beautiful.** Especially at crossroads, they are to be avoided. How to do that can be seen in Fig. 27, in contrast to the less desirable Fig. 26.

If a road splits into two branches, with one rising, and the other falling, then it is advisable to lay out the connecting section (pg and xy in Fig. 27) on the level, and then quickly let the road rise or fall for the first part more steeply than the percent slope of the road section as a whole requires. This little trick can be applied quite safely and with especially favorable effects in nearly level terrain, where the roads have a small gradient.

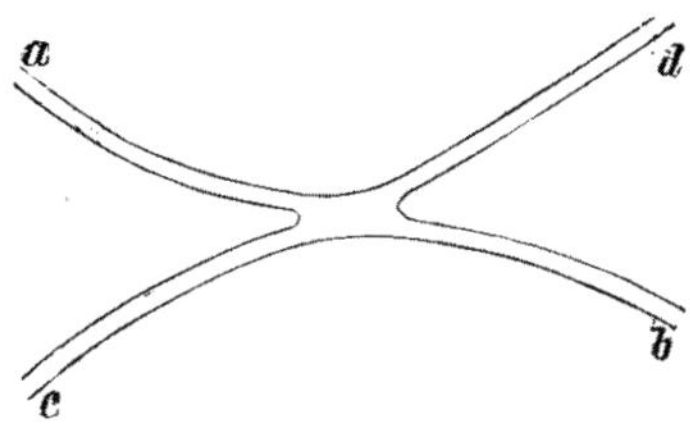

Fig. 26

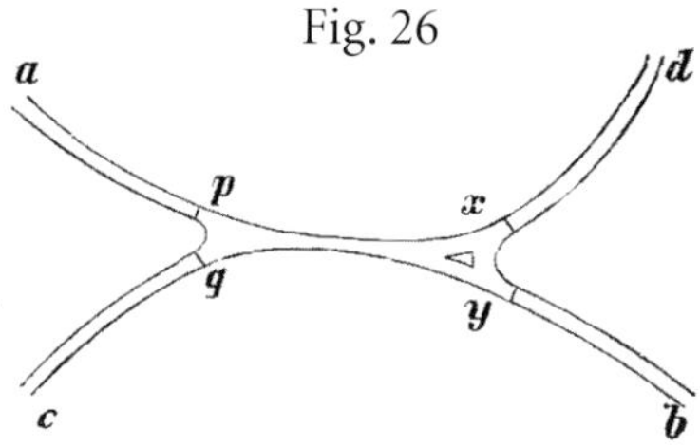

Fig. 27

It remains to be discussed which **size of administrative units** is more desirable from the aesthetic point of view, and I think in this respect that, ordinarily, the useful will also be the most beautiful. Silvicultural systems which create mainly even-aged stands of a single species will need **small compartments;** then a diversity, desirable in the aesthetic respect, will be achieved in that the cutting will not advance equally in the different management units (cutting a little earlier in the northern ones), but will bring the compartments of old and young trees together in ever differing spatial arrangements. Particularly with small compartments, the eye meets nice stairstep-like arrangements of the mature trees (looking to the southwest from the northeast). Such a view is shown in Figure 28.

Figure 28

Figure 29 shows in a schematic way how such views develop with regular sequence of cutting. Compartment 89 contains 100-year-old trees, already begun to be cut, starting on the east side. Compartment 116 is covered by older pole stands. Compartments 88 and 117 have some clearings and the remainder is young stands, so that one can still see over them. The directions of the arrows suggest that the views from the road meet quite different backgrounds and will offer changing scenes in a quick sequence.

The larger the compartments are, the fewer scenes will be able to be seen. But large compartments offer the aesthetic advantage that one needs to have, besides the firebreaks forming the boundaries, numerous other roads, which will make their interiors accessible. Many advantages of these roads have already been mentioned. Here I would like to add another one, being that these roads (being totally independent from the wind direction and the needs of the cutting system) **will be allowed to be located to the taste and the mood of the forest owner.** (Muehlhausen even suggests, under certain circumstances, to have the roads of the 4th order intentionally laid out into an inconvenient direction, so that the timber carter will be forced to leave those rather quickly; these roads are not constructed and laid out for heavy traffic.

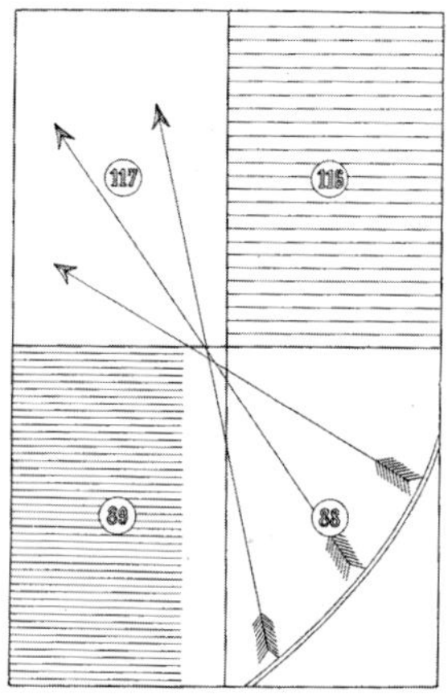

Fig. 29

If I were to subdivide the district shown in the example in Figures 24 and 25 into **only four** compartments, I would tend to solve the problem according to my taste (or my mood, as it may be called), about according to the pattern in Figure 30 (below).

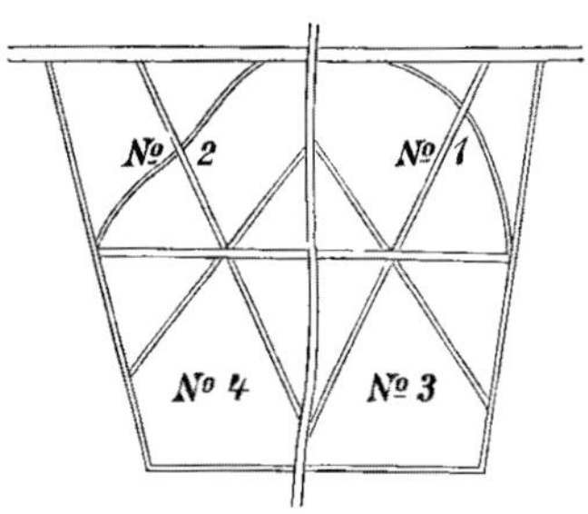

Fig. 30

Furthermore, it is an advantage of large compartments that a **vast connected stand of mature trees causes a much stronger impression than the same amount of mature trees frittered away in many pieces.** Where, therefore, the stands offer variety in themselves, and the barrenness of vast clearcut areas is excluded by the method of management, the compartments may be large. They are allowed to be, because hopefully the calamities which were to be prevented by the division of the compartments will disappear the more we tend to unite on one and the same area not only different tree species, but also different age classes which are integrated horizontally as well as vertically. The forest manager will perceive it to be a great benefit when he is going **to reach frequently a road, but rarely the boundary of another management unit,** for the data entry as well as the management are facilitated by that.

In regard to the **naming** of the evolved forms, it should be preferred from our point of view to unite **sub-district**s as a whole and to number continuously the management units within them, because by this an organizational structure is going to be achieved, which the forest district would not have had if the whole were divided directly into the small parts, each equal to the other.

Many practical advantages accrue to it to introduce names in addition to the unavoidable numbers, and even more aesthetic advantages. To justify that statement, I need to interject a general contemplation:

If some different, not really meaningful but pleasing impressions unite to form a total impression, they usually have a much greater effect altogether than they would have been able to cause individually and sequentially; in some cases one might be allowed to claim that they increase their value in the relation of a geometric progression. Let us for example think of the song: "Hail the man, who . . . the green wood." The thought which is expressed in it touches us sympathetically, but we would not stand without protest for someone to repeat in prose the content of the five verses to us on several successive evenings; the form of the poem is only moderate, and the good impression in content and form made by the first four lines has to help in neglecting the weaknesses of the rest. Nevertheless, **the whole** is suitable to offer a highly efficient foundation to the singing, for one will not be able to misjudge that this really is true and that the melody on its part is supported by the lyrics. With Italian lyrics, however nicely sounding syllables it might have had, this song would not have become established at our universities. On observations like that, Fechner bases his **"principle of aesthetic support or enhancement"** in his *Pre-education of Aesthetics,* and from this it is understandable that sometimes **good names in the forest** are so important. Good names are only those which essentially influence the impression which the site leaves on us, because they not only suit well, but also **inspire some other ideas.** These need to be related to the ones for which the specific site puts us into the mood. So, for example, the sacred halls near Tharand might not have been as famous if the stand had been called Buchenthal, not to mention totally unsuitable names as are often found. In the forest district Nesselgrund a firebreak, uncommonly long for mountainous situations, is called Ewigkeit [Eternity], and this is far better than to call it the long line.

It is the same with the names Paradise and God's Stairway in the head forest districts Katholisch-Hammer and Altenplathow.

Unfortunately the forest scenes which deserve names of that agreeable kind are not often found any more; but the principle is applicable to undesirable conditions, too; one will do better to name the worst area with pine soil of 5th class Grief Mountain than Dry Mountain.

Often there is a reason to draw an advantage **out of the previous history of a stand, out of hunting or any other strange events,** and often **dedication names** may get used. In these cases, too, the rule is valid, that the better the names, the richer ideas they inspire and maintain in us. Whoever wants, for example, in a forest district that once was directed by Burckhardt, to dedicate a subdistrict to the author of *Sowing and Planting,* will do better to name it "Burckhardt's Pleasure" than "Burckhardt's Mountain," for with the first name one can realize immediately: the man has been here, here he has worked and enjoyed creativity.

Head State Forester von Reuss gave a nice example of this type of naming when he called an exemplary plantation on an old burn in the Tucheler heathland "Schulze's Diligence" in honor of the forester working there.

In the head forest district Katholisch-Hammer, which is in my neighborhood, one pole stand is called "Pickel's Warmbeer." With that name, who can not visualize the honorable Forester Pickel staying in the plantation tirelessly in awful April weather, while his caring wife thinks about his health, keeping a warm beverage. If the site had been called "Pickel's Sowing," both the name and Forester Pickel would have been forgotten long ago.

It is not easy to immediately produce one hundred names or more like that where there is still a complete lack of fitting names. However, one will have less difficulty if one visualizes the abundance of relationships which one might use, like I am going to summarize here. One can relate

1. with the shape of the area and the boundaries
2. with the composition of the soil and the flora
3. with the stand and its history
4. with well-known things in the vicinity (rocks, trees, villages, castles, etc.)
5. with hunting events and other remarkable facts, as well as old tales.

Finally, to get a solution:

6. dedication names. These, however, if they are related arbitrarily, only have a chance to become established when they are mentioned and called often, therefore they are more suitable for sub-districts or longer roads and firebreaks than for management units.

The last Section [Section B] is going to discuss the outer marking of sub-districts, compartments, roads, etc., by numbers, inscriptions, road signs, road plantings, etc., because this is an excellent opportunity to decorate the forest according to the wealth and desire of the owner.

Chapter Three.

The Silvicultural Systems

As the establishment of the management units has to be coordinated with the design of the road network, so the choices of the **silvicultural system, the tree species, and the rotation period** stand in close relation to each other, but it will be possible to treat each of these choices in a chapter of its own **for the theoretical discussion.**

Primeval forest, in the strict sense of the word, cannot be the subject of forest business, for where management starts the primeval forest ends.

Nevertheless, a short contemplation has to be dedicated to the primeval forest, because several times it has been recommended in respect of beauty to maintain a piece of primeval forest or to allow it to develop anew, while one considers it as self-understood that the forester is going to be its guardian.

This, however, can be argued. The landscape gardener or the "woods-wise hunter" would probably be more suitable and effective. The economic management is in our blood too strongly!

The woods-wise hunter is going to be discovered among the earliest forestry people, in a development opposite to the one which once made the forestry people develop out of the woods-people.

In America it has been decided early enough to preserve a primeval forest because of aesthetic respects. The so-called Yellowstone Park contains an area larger than the Grand Duchy of Baden. In Germany it would take a long time before a plantation forest would be able to return to a primeval forest. A request of the Count of Tschirsky-Renard brought into the Upper Chamber in 1897 wanted to designate the Grunewald [near Berlin] as a primeval forest. The realization of his ideas would have been most interesting; nevertheless, I think it is right that the Upper Chamber did not approve.

Just the **risk** (falling trunks, breaking branches!) makes a primeval forest unsuitable to serve as a recreation area for large groups of the population. But this is not the only reservation.

A primeval forest, which would be such only in respect to the vegetation left alone to itself, would remain something really incomplete. Wildlife would need to enjoy the same freedom as the trees' growth does. But for this, even a forest ten times the size of the Grunewald would have been far too small. We Germans are not rich enough to allow ourselves the interesting luxury of a primeval forest; and after all, I do not know of any primeval forests still maintained anywhere in Germany, if one does not declare some of the barrier forests, maintaining themselves miserably along the borderline for tree growth, as such. Just recently I learned of an area of 200 acres in the Kingdom of Bohemia, the "*Lucken Primeval Forest,*" which is still maintained as a real primeval forest, but I could not get detailed information on it. A picture published in the *Gartenlaube* [Summer House] shows tourists in this forest climbing around on the rotten downed trees. The viewer will not be able to avoid the impression that primeval forest and tourists do not go together well.

The so-called **Neuenburger Urwald** in the Grand Duchy of Oldenburg, situated close to the railway line from Oldenburg to Wilhelmshaven, is not a primeval forest in the original sense but a grazing forest.

To have maintained the **grazing forest,** this old German form of cultivation, up to modern times, at least by this well-chosen example, remains a continuing benefit provided by the highly blessed Grand Duke Peter of Oldenburg.

In prehistory, our ancestors in general used the forest nearly exclusively as a pasture for their cattle and for the satisfaction of the local need for fire-wood, in remote areas up to end of the 18th century. The axe did not dare to come close to forest giants hard to cut and even harder to process. The old trees with wide canopies, as long as they were beeches or oaks, were saved for their mast. So the forest scenes evolved, of which Pliny has left behind such a graphic description. I was able to admire a pale reflection of such magnificence near Postel, as old Count August of Maltzan still kept in the Walkaw Forest a piece of pasture forest, as an "estate luxury," as he used to say jokingly.

The "Neuenburger Primeval Forest" mentioned above, with a size of about 50 ha, is part of the Forest District Neuenburgerholz. The surrounding villages exercise a right for grazing therein. The stand is formed by oaks with wide canopies and some beautiful beeches. In the understory, hornbeam, soft woods, hawthornes and holly *(Ilex)* exist. The very beautiful hollies have been suppressed during the last years by the hornbeams, and so these had to be partly cut.

Old trees are not cut; the falling trunks are left lying if they do not bar the way.

The **plenter forest** [selection forest] has been glorified in literature up to recent times as the most beautiful type of forest management. So, for example, Quaet-faslem, when he said to the North-West German Forest Society: "The lot-wise, or group-wise plentering, which permits group-selection as well as single-tree selection, which removes poorly growing parts of the stand and regenerates again according to the shade tolerance of the individual tree species, be it natural or artificial, allows the most consideration of all demands of landscape beauty, variety between deciduous and coniferous trees, shade, introduction of tree species, preferential treatment and care of individual beautiful trees, and allows the specified part of the stand to always stay forested, and for views of younger stands to intermix gracefully with older ones on a small area."

"It is adaptable in respect to opening, or maintenance of vistas and free framing of landscape scenes, and offers the greatest possibility to weld nature and art together imperceptibly."

At that time the speaker thought of conditions like those found on Kluet Mountain near Hamelin or in the Eilenriede near Hannover. There selection management is quite suitable. On a **large** scale it does not suit aesthetically nor economically. Recently von Guttenberg has proved that in general the **high forest** should be preferred.

The magnificent splendor of regular high canopy forest stands has been described by Schleiden so accurately that I cannot resist quoting his remarks here:

"If we wander through the Harz from Wernigerode to Ilfeld and turn, before we reach it, from the street up to the forest house of the Count's Forest District of Wernigerode and Sophienhof, so a nice man and competent forester, the Head Forester Kallmeier (if he is still alive, for it has been a long time since we were there), will lead us into his favorite stand, a magnificent vast section of beech high forest. The smooth, whitish-grey columns of the 120–125 year-old trunks, 18 to 24 inches in average diameter, rise upwards to fifty and sixty feet without branches, carrying above a dense dark-green dome which does not allow a single sun beam to invade; the ground is covered by a dense even carpet of old brown leaves, not broken through by any plant growing. So this magnificent dome is standing in **silent majesty, and among its forest of columns the small human disappears** as an unimportant appearance. Or let us go from Fredelsloh through the Grubenhagen Mountains to Reliehhausen in the Hannoverian Solling, so we will pass through a hilly forested area. Walking along on its short lawn, decorated with friendly little flowers, we are surrounded for hours by marvelous stone oaks [*Quercus robur*], many hundreds of years old, each tree with a strong trunk and rich crown, separated from its neighbor by a little space, so that the whole creates an impression of solemn quietness, but also of **sunny cheerfulness.**

Or let us again change place and time. On a fresh morning in late fall we walk through a mature stand of spruce in the Thuringian Forest. The soft elastic moss rarely leaves space for another small plant. The slim, parallel trunks rise up to 80 feet, and the heavily-plaited tops form a dense roof that has prevented any sunbeam from reaching the ground for half a century."

I suspect that those speakers praising the selection forest, who prefer that management type totally and unrestrictedly, would arrive at a less favorable attitude if they had taken an entirely balanced point of view. I got the impression that the defenders of selection management had created in their minds eye the fantasy of this management type as being guided very carefully according to all the rules of science and practice, in order to confront it soon with a schematic, ruthless and unmindful management by clearcutting, a management type which has accurately been called the razor system or destruction method.

With the same justification, one might compare on one side the high forest in all its different, rich variations, which it owes to the loving care of attentive guardians, that means with understory and canopy, with a friendly mixture of tree species with pleasant variety, like the different types of regeneration offer it, which takes place sometimes on a larger area, sometimes on a smaller one, often as advanced regeneration on the smallest area, sometimes in an open situation, sometimes under a canopy, every time accompanied by a characteristic pile of graceful grasses and herbs, at dawn and dusk brightened up by the delicate roe, the cautious deer—this being one picture—and on the other side a forest with short, branched, low-value mature woods, with crippled young trees and meadow-like clearings, everywhere carrying the traces of thoughtless cutting, careless timber transport, and all this an infinite monotony which never allows the eye to roam into the distance.

Fortunately the **high forest** on its side does not need its defendants to fight for it with such unequal weapons, especially if its rotation period has not been set too short, and even in this case still much can be performed by shelterwood in groups or even as individuals.

When I worked on the first edition of this book, I often rode horseback out to two stands which had been kept since the time of the unregulated selection forest in the vicinity of my place, namely the "Paradise" in the Royal Head Forestry Katholisch-Hammer and into a lot of the Gontkowitz Forest, belonging to the rule of Militsch. The Paradise was formed mainly by beeches with a lot of in-mixed oaks and pines, and the Gontkowitz Forest by pines with spruces. Both sites showed incomparable beauty and I already thought I would be forced to take back what I had written in favor of the high forest, but after thinking about it I had to tell myself: These stands contain a treasure of ancient woods and they owe to this advantage the power of impressions they make on us, and maybe as well to the circumstance that they are the only ones of their kind for square miles around. If we had the opportunity to see a high forest whose average age was almost 100 years older than that of the surrounding forest in the midst of vast younger stands of plenter forest, certainly there it would captivate us even more than does the other one here.

By the way, I do not intend to deny at all that the **selection forest and the silvicultural types related to it are characterized by the abundance of different and richly varied details, and that therefore they offer a substitute for the imposing development of the magnificent domes of the high forest, or that they even can outstrip it on the right site. This will be the case everywhere where the situations force one to do without mass effects,** first of all in an area which tears apart the unity of the stands by brusque and quickly changing relief and catches our attention so much that the forest is to be regarded as, so to say, decoration. Furthermore, one will feel obligated in **small areas** to take advantage of the rich changes in detail and the less clear openness of the selection forest, and one might possibly decide to manage the **edge of greater**

forests in a selection-like manner on sites close to cities or recreational towns; for their inhabitants usually have to restrict their visits to the nearby area due to lack of time and physical ability, and what exists beyond that is of no use to them.

In hilly regions, however, a special recommending advantage favors high forest system, so that we are only allowed to leave it really exceptionally even with conditions existing like the ones described above. For if one carries out clearcuttings with standards here with some consideration of the **view from the hills,** one is going to gain each year the most graceful views and the nicest scenes. While one vista becomes overgrown, one already begins to open up the second one, and one always enjoys new designs. I might be allowed here to quote a part out of one of Prince Pueckler's letters, as it has been handed down by Petzold in his *Theory of Color:* "In respect to the theory of colors I have also had an experience this winter. You may remember that in front of the windows where I live, the horizon was limited closely by a pine high forest, a complete curtain of **one** height and of **one** color. I have given now a very picturesque zig-zag horizon to the pine stand, and also quite different colors, by cutting about 500 cords out of the forest. The front groups now stand up in black-green, the more distant ones appear light green and the far away ones, having become visible just now, present themselves in different shades of blue. Quite an art-like shading , and still it is one and the same low pine forest, no tree in it more than 40–50' of length at a maximum, nearby, and all of the same color."

Management types that keep the whole area about evenly stocked with more or less old wood for decades cannot offer such change, such surprises. This is different in the mountains, where one will often need nothing else to do, in order to open up a vista, than to cut one branch from the first tree and cut down the next lower tree. The third one in the row then can be left there, because our eye will already look above it; it is not only allowed to, but has to stay, because otherwise the scene would lack a foreground. Fortunately such steep slopes are the normal place for **selection forests** anyway.

An especially appreciable aesthetic type of the high forest is attained by **clearcutting with standards** [similar to seedtree cutting]. The shorter the rotation, the more important is the maintenance of trees with a promising future into the next rotation. The area saved by increasing a short rotation for several years is hardly noticeable in the beginning, while leaving seed trees, even if only some percent of the previous allowable cut, is soon going to change the entire character of the district. This change, however, will only be for more beauty, if clever preparation during many years has initiated the action.

As the saving of seed trees turns out too often to be an economic mistake, it can miss its purpose aesthetically, too, if it is done in the wrong way. Pines turned crooked, beeches suffering from sunscald, oaks with top die-back, may be picturesque sometimes, but they cannot satisfy us in regular forest management. Very early, that means at the occasion of the last thinning, 15-20 years before the cutting of the stand, the release of the trees to be saved must be done. At the same time their lower boles have to be screened by understory vegetation.

One might want to visualize more often the high aesthetic value of a magnificent seed tree; then one would not think of the attention, which is necessary for its selection and preparation for being left as a solitary tree, as being too much effort; and what an abundance of economic advantages one would have harvested! Here I may be allowed, in an exception, to go into detail and to list the economic advantages of seed tree silviculture:

1. Good seed tree trunks produce the highest values with relatively low capital input.
2. They show the capacity of the site.

V.
Postel. Ueberhaltsstämme in natürlich verjüngter Kieferndickung.

3. They serve the establishment of mixed woods on the most suitable sites by seeds carried far away.

In this respect Danckelmann writes: "As already mentioned, in the regeneration of the mixed pine stands, grown up out of selection forest, partly **beech and hornbeam poles and inferior species were saved individually and in groups as seed trees.** These have created, after having become capable of producing seed, a dense ground-sheltering understory, far beyond the area of their crowns, below the thin crowned pines. The stands grown out of this have become the example for the pine-understory-management established here."

From my side I can add that also here in Postel in several pine pole stands, single older beeches, surrounded by rich seedlings, decorate a forest area very well, which otherwise would be barren and empty.

4. They make orientation in the forest district easier, which may be particularly useful in observing from a higher point (forest fire, surveying).
5. With seed-bearing species, the very old seed trees are often the only trees in the forest which offer acorns and beechnuts to the forester for his sowing field, and to wildlife as winter food.
6. Also in some other respect the shelter-trees are useful for hunting while facilitating the shooting of birds of prey, further on as courtship trees, as a cover for the leaned-on hunters.

Quite generally it is accepted that it is criminal to cut old forest giants. Just like their maintenance, one should care for their replacement, and this can be achieved only by shelter trees.

With this, one should think not only of the "1,000 year old" oaks. More modest tree species also have a right of equal provision.

"Why did you save that pine over there?" Dr. R. asked the Royal Head Forester in X. "Because it is worth 10,000 marks," was the answer. "How is that?" "Quite simple. An artist painted the tree last year and pretty soon he received 10,000 marks for the painting."

One should not argue that the most beautiful forms for seed trees can be raised only in free position from youth on. This widespread attitude is not correct. The greatest expert in this field is Prince Pueckler, who planted a thicket at many places where a light grove was to be established, because he believed that the **most perfect tree shapes grow up where the young growth happens in a closed stand.** However, the oak seed trees of the high forest, often with top die-back, and the pines, turned lopsided by the wind, which eke out a deeply melancholy existence, scattered as single individuals over many cutting areas of the Mark, appear as a proof of the opposite. But how different would these trees look if they had been taken into preparation treatment only two decades before their release by gradual thinning and understory development in their closest vicinity. For this, only few groups in the area that is in the early establishment period need to get the care, which defenders of selection management have to presuppose as possible for the whole forest, if they do not want to grow picturesque but unutilizable trunks.

If we now turn back to the conditions of hilly regions, then we also have to remember the **composite forest management [coppice with standards].** Whoever has frequently seen this in such regions will probably remember how much the eye likes to roam downwards along the mountain slope between the two leaf roofs, below the crowns of the high canopy and above the fresh green of the young shoots from tree stumps. There, one also has the opportunity to do without part of the high canopy and to gain valuable vistas by this small sacrifice, a goal which one finds often aimed at by cutting out the tops of trees in the most miserable way conceivable.

If Schrember, an advocate of the composite forest, praises it for being the "aesthetically most beautiful forest type" "to experts," then he has seen the landscape of the foothills of the mountains in front of his eyes, but the composite forest also has its advantages in the plain.

Riding on the pond dams of the local region here, which is rich with fish, one can let one's eye roam around just like from a mountain, above the shoots and below the main canopy.

At the same time, one enjoys the great diversity of the tree species. The soft deciduous trees, otherwise often tiresome, are here as a canopy wood in their right place, and give a friendly look to the landscape. The same feeling invaded me, when I once traveled from the serious beech forests of the "Holsteinian Switzerland" (area near Eutin) to the composite forests of the lower valley of the Weistriss. Besides the diversity of the tree species, it is also the irregular canopy that, in contrast to the densely-closed leaf walls of the beech stands, causes the more friendly expression.

While studying Burckhardt, my attention was drawn to the fact that composite forests show especially rich and diverse ground cover. In the bottomland composite forest, I often had admired this advantage (snowdrops, cowslips etc.) myself, but I had looked for the reason in the quality of the soil. Later, detailed observation taught me that equally favorable soil in high forest management does not even approximately appear that rich with flowers.

As the selection forest is mostly overrated, **the low forest** usually is not appreciated enough. The **fresh germinating power** intrinsic in the shoots from the tree stumps always offers a lot of pleasant things to the viewer (the lushness of a chestnut low forest once deeply impressed me, although I only hurried past it); its main attraction, however, lies in the **beauty of the winter colors.** By the warm shades of the young shoots' bark, the stump sprouts become really suitable to offer an attractive sight in the cold winter landscape, much more than conifers are able to do, for the latter easily appear too gloomy in winter; seen from some distance, they are almost melancholy, if they are not decorated by ice and snow hangings. Young stump shoots usually keep some brownish foliage until far into the winter on their branches; this is a particular advantage of the oaks, which offer a rich substitute for the cold grey-brown of the bark of its young trunks and branches. Beside such winter-foliated shoots, the conifers are not in danger of appearing too gloomy, because their green gains a pleasant liveliness by the contrast with the brownish yellow of the foliage. Therefore, in this point, the hints of aesthetics come together well with the needs of practice, for the latter likes to plant a group of spruce at sites with shallow soil in the low forest. Inasmuch as the shallow soils occur mostly on the projecting heads and ridges, this procedure serves very well to bring the shape of the relief to a positive appearance by clearly stressing its characteristic points and lines. So the low forest may appear friendly and even as an adornment of a region, but it will never be able to give a magnificent impression. That is why it is to be bewailed that particularly on the banks of our River Rhine the **oak peeling forest** [coppiced to produce tanbark] appears dominant and wide spread as a forest type not really fitting to the whole. It is hardly more favorable than the nearby vineyards, which are also able to satisfy aesthetically only when supported by idea-connections. Idea-connections are also associated with the oak peeling forest, but these are unpleasant ones, at least there where the poles are peeled while standing. For rather unwillingly and almost totally unconsciously, but therefore the more emphatically, we put the tree's growth on the scale of human living relations. So we pity the young vivid shoots, which become a victim of the axe when they have barely reached the border of fully vigorous growth, as if it made a difference to the tree whether its destiny is going to catch it in youth or in old age. Although the awareness is comforting to us that at this age at least the bark is ripe, if not the tree; but if we even see the young oaks, peeled on the standing wood, standing there naked for weeks, then we associate this procedure with the feeling of being tortured alive. I have heard several times that inhabitants of the eastern provinces, not being used to this sight, have been upset for years over a peeled tanbark hedge that they had seen. However, it cannot be demanded that one is going to do without this method at all; but near heavily visited recreation places and along country roads one might remember that the urban citizen, escaping out of the narrowness of the streets into the outdoors, brings with him sensitive nerves for things like that, and one might want to pay some consideration to that.

The same is true for the **lopping system of management.** If we see a trunk rich with scars standing there bald, recently roughly de-branched, in the middle of summer, then we think we are to bewail it as having been contemptibly abused. **Therefore one would not like to find lopping wood management anymore along roads; in the lopping system small bushes or as a canopy wood the so-called deciduous trees are often really suitable,** and they may even look quite magnificent there, like the Pyramid Poplars. In some ways even prettier than these are the lindens, whose praise, of course, relates only to the clipping of previous years. Therefore one should not **clip all of them at the same time,** but each year only about a quarter of the trees. With this division, **those newly de-branched disappear** among the others. Anyway they are hardly noticeable from the distance and therefore are not able to seriously disfigure the landscape. If handled in this way, the lopping system of management suits quite well for small estates, and one always may be quite glad when at least this form of wood use maintains the last remnants of forest for a region, as is often the case in Silesia.

The small picture [Figure 31] shows such a management type.

According to experience, lopped trees develop very picturesque forms over time, if they are left alone.

Pollarded trees become picturesque with age, too. How much poetry lies there in old pollarded willows, whose rugged, often overgrown, repeatedly violated and always revitalized trunks take on highly strange forms, and decorate their old garment with an often rich vegetation, in offering habitat to birches, *Sorbus, Solanum,* ferns and many other plants.

VI.
Postel. Alte Kopweide.

Fig. 31

Whoever still wants to plant willows for pollarding should pay attention to the beautiful warm color of the winter's bald switches when choosing the cuttings for planting.

Three beautiful red-brown pollarded willows can possibly decorate the landscape better than a large group of oaks would be able to.

Quite according to the teaching of Paragraph Two, Chapter Two, Part I A, saying that the right adornment is always supposed to show something, Gilpin recommends the planting of trees for pollarding only where it is intended "to indicate a swampy area, or, in a medium elevation the winding banks of a river, creeping along lazily in its deep bed, which cannot be distinguished in another way." This is indeed a sign language generally understood and important in the country of fox hunts.

Chapter Four.

Choice of the Tree Species.

A stand growing vigorously is more beautiful than a less flourishing one. That is why the aesthetician as well as the rationally calculating man of practice will prefer tree species which will grow best on the actual site. The forest aesthetician is distinguished from the enthusiastic layman by his forward-looking view. He cannot be pleased with pure birch on a pine soil of Class III, however good they are growing in the beginning; for he sees in the future that very soon the growth will stop and the soil is going to become impoverished. Similarly, we judge pines on a site known for snow-break, or on the best oak soil, where they are going to produce only light, worthless timber. Similarly, too, spruce should not be encouraged on sites where it does not belong.

Whoever practices forest management as an art will manage according to the potentials of the sites, but will present himself as a master, particularly in light of this limitation, in avoiding any sort of monotony.

Once, the pure beech stand was held up as the ideal goal, and unfortunately it has too often been achieved. Nowadays there are conifer fanatics who are equally eager. I know a forest owner to whom any sycamore maple sprinkled into a spruce stand is an abomination. However, I cannot hope for approval by people like this if I claim that deciduous tree stands are more beautiful than coniferous trees, mixed ones more beautiful than pure ones. This is supposed to be valid only in general, for not even I am going to prefer a heathered birch grove to an unspoiled fir stand. Examining for each special case which tree species should be chosen, one should **prefer in doubtful cases the tree species which is more rare in the region.** If one possesses in a vast pine forest some hectares of better soil, which carries beeches at best, one should not let the opportunity be wasted to establish there a deciduous tree stand. One day the descendants will like to visit this place for joyful rest in the forest; on the other hand, the forester busying himself under the darkly shading beeches will breathe more freely anytime he walks through the light pines to which an ancestor gave wisely the hill of lighter soil. He is not going to drive out the more modest tree species from its place.

The aesthetic qualities of the individual tree species have already been discussed in detail in PART I, B, Chapter Four.

Combining into complete stands the shade tolerant tree species (fir, spruce and beech) on the one hand, and the **intolerant tree species** (pine, larch, oak, softwoods) on the other, forms contrasts.

The tolerant species do not permit understory in the long run, and this may be an advantage or a failure, according to the circumstances. It is an advantage where the terrain bears beauties that should not be hidden (rocks, ravines, picturesque variations from slopes to rounded hilltops); it is a mistake where the area is without attraction. Wilbrand's description fits such conditions:

"Even the usual shelter-selection management of beech, the ideal of many forest lovers, has to be criticized by the aesthetician. This includes the point that this management, in which, due to the use of the heavy seed years, evenaged stands grow naturally one after one another for a longer distance, has its weaknesses in respect of beauty. Under a closed beech canopy, no other wood species is going to grow, nor hardly any other plant. Therefore the ground itself is without vegetation. It is covered with the pale foliage of the previous year. However excellently this is working in regard to forest management, and however much we need to manage in the way that we are going to give and maintain a really thick cover of this precious material to our forest soil, the cover itself surely is not particularly beautiful. The green canopy of the older beech pole stand is high up in the air, but down on the ground there are only pale leaves; the eye does not see enough green; in order to see such green, one has to tilt one's head far back, and this is not really pleasant, as everybody knows who has ever marked trees for cutting."

How this undesirable result of shelter-selection management of beech can be solved Wilbrand himself has already reported. It needs only some advanced reproduction groups in order to brighten up such stands. Plantings of young oak trees in fair sized openings will produce the desired results the fastest.

As seen from the distance or from above, the tree species are to be grouped in another way, namely into those with **pointed** and those with **rounded** tops. Spruce and larch belong to the first group, beech and oak to the latter. Fir and pine belong to the first one in their youth, to the latter one in old age.

The landscape gardeners agree that trees with pointed tops fit excellently near buildings and into landscapes which develop their shapes in straight, preferably horizontal lines, as in Fig. 32 (spruces in front of a mountain ridge).

Fig. 32

The cypress is really the right tree for the neighborhood of the Greek temples, and with us, one likes to interrupt the sight of long-stretched barn roofs by a group of Pyramid Poplars planted in front of them. **On the other hand, sharp-angled buildings demand round-top trees,** and that is why long ago the lady of the Knight Toggenburg chose a monastery, which "shone out of the middle of dark **linden.**" Conifers do not fit at all to the sharp gables of medieval buildings, except for very old pine trees, grown up in full sun.

Hidden in our forests, numerous ruins of magnificent old buildings are entrusted to the forester's loving care. Thinking of that I could not avoid these remarks borrowed from the art of gardening. The stated principle is exemplified very aptly with the Castle Groeditz. While in the immediate vicinity of the main building there is an abundance of linden and maples, the western enclosure wall, running horizontally for a longer distance, is towered over by the tops of some old firs. Rising out of the low forest, these interrupt the monotonous line of the wall if one looks at it from below in a way as favorable as they form, the other way around, seen from the castle, a very suitable foreground in front of the horizon, running rather flat into this direction.

Whether **pure or mixed stands** are more beautiful cannot be decided easily with any degree of validity.

Pure stands give a sublime impression in old age by the magnificent enfolding of homogenous shapes; mixed stands are more friendly and more inspiring because of the variety. But both advantages can be united by a very well chosen, very moderate embeddings of mixed species into pure stands. One should improve their appearance by a **very moderate embedding of other tree species, up to about five percent of the individuals.** For adornment this will always work well, without disturbing the desired mass effect. The tree species forming the stand will keep its dominant character, especially if the added species themselves have a diversity of character. For example a pine stand, if it has been mixed with 5% beech, will not appear any more attractive in May and in fall than a pure stand, but it will do so, if aspen, birch and oak share the 5% with the beech. By the way, these statements relate only to older stands; plantations and pole stands have to show richer embedding, if the desired success is to be saved for later periods.

Concerning the **mixed stands** in the traditional sense, it is not to be done with the simple even mixing of different tree species. For example, one hundred hectares, stocked with spruce and pine in totally even distribution, is not going to have any advantage over an equal area which is stocked on one half with pure pines and on the other half with pure spruce; quite the contrary, the latter relation certainly is the one more attractive, due to the contrast. If, however, pine and spruce are united with each other in such a way, that first the one, then the other species will appear more often, and each one will express visibly the special character of any site by nearly dominating in large groups locally; the spruce will allow itself to cover every now and then as an understory the base of pure pine groups, while not far away, storm-resistant pine forest veterans tower occasionally over closed spruce pole stands: then we find a reasonably interesting and beautiful mixture, rich with variety and change. It can receive a further adornment by the addition of about one-and-one-half to two percent of deciduous trees or larches. The same can be applied with the same objective to the mixture of oak and beech, of alder and birch, etc., also of pine and sessile oak. Except for this latter combination, I am not taken very much with the mixing of deciduous and coniferous trees, because one loses with it the gain achievable by the stronger contrast. I recommend **letting mixed deciduous stands alternate with mixed conifer stands with each part embellished by the embedding of conifers or deciduous trees up to five percent, as mentioned above.**

Groupwise mixtures are to be preferred as a rule to the **stripwise** ones; the latter ones appear less natural, but they also have a certain attraction. If they have been established with accuracy, they will arouse this pleasure, which order, symmetry, and neatness are able to gain at any circumstances. However, this pleasure will be reverted if the planting does not grow, for example the rows of young trees do not want to grow and the soil below them is becoming poor and hard.

I have to defend myself explicitly against the assumption that I had intended to establish strictly-to-be-followed rules in this chapter; for I am quite aware that sites, management types and other conditions can influence the decisions in so many ways, that **all advice not limited to a certain single case can only be taken as an approximate orientation.** This reservation, which is valid particularly for the chapter just finished, may be related, to this whole second part of the *Forest Aesthetics,* with little limitation.

Chapter Five.

The Determination of the Rotation Period.

Under aesthetic points of view it is not hard to determine a lower and a higher limit of the length of the rotation period. The lower one can be found in the words of the well-known and popular song, which is often sung: "Who has **built** you up, you beautiful forest?" etc. According to that we should allow the trees in the high forest get at least high enough that it can be said, with some justification, that the forest is "built up," that it contains its columns, which united, form the wall of wood at the edge and support the canopy of crowns.

The **upper limit of the rotation period,** however, is to be regarded as already exceeded when the economic decline becomes conspicuous. **Therefore it is not justified in the forest to move declining stands into later periods.** They are always picturesque, but they conflict too much with our concepts of utility to allow us to think of them as beautiful in reality. We are going to end the rotation and establish a new stand, and set the rotation period so carefully that similar pictures are not going to develop again.

This subject has been treated very nicely by von Wildungen. (in "*Waidmanns Feierabend*" ["Hunter's Evening"] 1815): The brave forester has met two individuals in the forest and has addressed them sharply. Now the following conversation follows:

"Painter: Oh, allow yourself to become calm, dear forest god! I swear by the sacred shades of a Raphael, Titian and Michelangelo, that we are not dangerous tramps, but traveling geniuses. Tired, we took a rest here and were delighted by the indescribable beauty of this forest, the true picturesqueness of its arrangements, its marvelous highlights.

Forester: What, man, are you even crazier than your chum? I think you even want to mock me! Is this lot supposed to be thinned out even more? Is it my fault that my ignorant predecessor had all his lots thinned out too much, that here it has been sinned so shamefully, and that all the time all leaves have been scraped away? Then the few remaining old seed trees had to die before their time! Now, however, there is no rescue any **more** for this cursed district—unfortunately—than to cut clear everything and to sow pines! And you dare to praise this ugliest blot in my forest, where for twenty years no hare has been able to hide, in order to mock me? Come on! Now I do not want to hear anything else any more! If you do not leave immediately, I shall whistle for my forest assistants, who are nearby, and shall have you arrested for vagrancy!

Poet (jumping up to his feet, to the painter in a low voice): Come, darling brother, let us walk away from here! Whoever is not enchanted by such a magnificent forest, who can dare to threaten marvelous trees like these with the axe, has—by Apollo—no aesthetic emotion, and what is it we might not fear from a barbarian like that?

Painter (picking up his walking stick and sighing deeply): Nevertheless it is sad that our superior picturesque and poetic attitudes always come to collide with rough people! Lucky me, that this morose forest devil, however, cannot prevent me from painting this unforgettable forest as beautiful as it appeared to **us.** Our expulsion out of this paradise is going to provide a couple of wonderful figures for that .

Poet: Also I want to sing about it in a passionate ode, like certainly none has been composed since the creation of the woods. May it please the forestry-minded fault-finder or not—**we** liked it and we were glad in it. Happy illusion is after all the element of poets and painters, and the golden teaching:

A delusion which makes us happy
is equal to a truth which presses us to the ground.

is the highest and most noble philosophy."

While the greater part of the public may be inclined to agree with the poet, we foresters feel, together with the old forester, the fact that a considerable treasure, collected over many generations, in mature timber and in the richness of the soil, perishes is a "truth which presses us down to the ground." We cannot look at scenes like these impartially, and therefore we cannot not think of them as beautiful, even if we wished to. The art of forestry, as an art which deals with reality, is not allowed to borrow its rules of beauty from the artists, who live only on the beautiful appearance. With the same justification, could the statesman otherwise want to order the relations of public life and of the family also in a way that they are going to offer material to the poet; but he would hardly find approval if he tried to put into effect social and domestic conditions like Schiller may have thought of when writing his "*Gang nach dem Eisenhammer.*"

The art next to us, the art of gardening, is in the same situation as we are, for even it has to remember that "the pleasure of the eyes is less charming than an unpleasant sight is disadvantageous."

After all, we have to accept this limitation, and we **have to comfort ourselves that we have some advantages in our field in comparison to the realm of appearance;** for example, **"occasionally utility can substitute for the lack of beauty in a real event, but never in a painting."**

Between the easy-to-prove upper and lower limits, there remains a wide range which might lie between 80 and 200 years with pines, and between 80 and 300 years with oaks, in well-managed forest. Now, how to find the right number?

Here it is appropriate to deal with the **theory of the highest net revenue,** which is known to claim obeisance for all the forestry issues, for example choice of the tree species, but deals mostly with the determination of the rotation period.

Overeager disciples of this teaching wish to forbid the aesthetician to join the discussion of this topic. Different are the masters. Pressler, Judeich, Neumeister and many others testify that the theory of the highest net revenue does not necessarily conflict with the care of the beautiful in the forest, even in regard to the determination of the rotation period. Pressler himself has said that his forestry is going to be "aesthetically very commendable, too." At the same time, he has pointed out on the first page of his *Forest Establishment* that the "granting of individual pleasure" can be accounted for as a revenue of the forest, if one is going to define broadly the term net revenue.

While Judeich, remarkably, mentions this limited concession only in fine print (the 2nd edition is available to me), Neumeister expresses himself more accommodatingly: "Even the most meticulous financial management allows that the influence of the forest on people and the country is brought into the equation, and even for the financial policy the imponderabilia are no *terra incognita.*" Neumeister, however, has stayed far behind Wilbrand, who stated as early as 1893:

"Especially interesting are the relations of forest aesthetics to the calculation of forest value and to statics. The prospering of many settlements is essentially dependent on whether the forest in its neighborhood is maintained as a beautiful landscape. In regard to the forests which lie directly near towns which are visited by tourists in summer, this claim hardly needs to be further explained. The number of such towns is large and it is still increasing rapidly. Some examples of the numerous places of this kind are in the Vogesen, in the Hardt, the Schwarzwald, the Rauhen Alp, the Taunus, the Odenwald, the Harz, the Riesengebirge, etc. Imagine that a beautifully kept private forest near a well-visited spa or climatic health resort is to be sold. The owner of the spa hotel will hardly allow it to happen that this forest is going to be bought by a timber merchant who is going to cut it down and to create a desert, for the cutting of the forest might mean as well the devaluation of the hotel. Accordingly, the owner of the hotel is going to offer more than the timber merchant, or than a competitor of the landlord, in whose interest it might be to deprive this place of its main attraction. So the financial value of a forest can be increased considerably by its aesthetic importance."

"But the maintenance of nearby forests is not only of considerable importance to the prosperity of individual hotels and smaller towns, but especially for some even larger cities. Numerous prosperous families change their place of living. The officer who has learned to know many garrisons of the Empire, chooses a city, when he is going to retire, which is the best able to satisfy his demands; similarly, the civil servant, and the business man who has become a pensioner can spend his interest money wherever he wants to. The attractions which have a particularly strong effect on many of these welcome settlers include nearby well-kept forests with nice trails for walking. Some cities have to be counted in this category, for example Frankfurt am Main, Darmstadt, Wiesbaden, Freiburg im Breisgau, and Eisenach. Every prosperous family which moves into such a city gives cause that a piece of agricultural land is exploited highly profitably as a building lot, and with every new necessary construction, a greater capital is settled down in the city, the value of the property in the downtown area rises, businessmen, bakers, butchers, merchants of all types find their income, and the state and the community gains cash in form of taxes or rentings."

It is founded on a correct appreciation of these relations, when Wilbrand further states: "Each single formulation in the management plan has to be examined in detail in regard to its effect on the current and future beauty of the commercial forest."

If now adherents of the theory of the highest net revenue think it is appropriate to calculate the aesthetic values of the forests together with the other profits, then one misses the scientific extension of the theory into this direction. Wilbrand admits quite openly that it is alarming, if, with forests of an outstanding value of beauty, one wants to solve important issues of statics by formulas. I would like to quote his thoughts: "Admittedly the solution of static problems is made considerably more difficult by this, because it is very difficult, nearly impossible, to express in numbers those very real effects of aesthetic actions mentioned above on the cashbox of the forest owner. We have to comfort ourselves with the thought that often, particularly with the largest actions of the state and the communities, for example railway construction, it is not possible to forecast the effect in numbers, and that very often greater emphasis is to be put on the indirect effects than the direct income which can be expected out of the project. We see, in many cases in forest value calculation and statics, it has to be calculated with indirect values, for which arrangement the formulas do not offer any instruction and maybe cannot offer any at all, except for the forestry figures, for which our formulas show the corresponding places where they have to be put into. At the same time especially the supporter of the theory of the highest net revenue profit is going to be the least opposed to having these indirect values being considered in an appropriate way. After all, it would contradict the very nature of this purposeful method if he were to reject it. For we will never succeed in expressing correctly in numbers those actual, financially most important factors; we end up with the result that it is doubtful to want to solve issues of the statics for forests of the type discussed above only with the use of the existing formulas."

I think it is unlikely that the difficulties accurately described by Wilbrand have prevented Pressler, Judeich and others from including the contribution of beauty of the forest into their formulas and calculation examples. They have often overcome greater difficulties when they introduced future timber prices and other quite indeterminate quantities into the calculation.

Probably the reason lies deeper, and the failure can be traced to the qualities of human nature: For it is an often repeated perception that many people act too rigidly and ruthlessly just because they are afraid, being aware of having a warm and soft heart, that these qualities might gain too great an influence on their actions, and be too detrimental to their own and other people's interest. It seems to me that this is happening especially often with us foresters. We have chosen our career out of love for the forest, namely love for the forest as an embodiment of ideal goods, but by no means because it would have appeared particularly attractive to us to supply our contemporaries and our posterity with timber. If now the theory of our chosen profession really becomes practice, so we cannot doubt anymore that our primary task is to grow timber, as much as possible, as good as possible, and at the least possible cost. Then a change takes place with the young forest enthusiast (sooner with one, more slowly with the other.) He discards completely his first ideals, and entrenches his mind artificially against their return; this is quite unreasonable.

In this sense Pfeil complains, certainly out of personal experience: "The material need allows less and less attention to be given to the sense of the beautiful in the forest economy. First the magnificent large old trees disappear, then the single picturesque group of trees, and finally, the monotonous deathly grey pine drives out the friendly, colorful deciduous trees. The same development which happens with the change from the poetry of the forester's life to the dull eking-out of a miserable existence in the factory, where the human is only regarded as a part of the machine, can be observed everywhere else. **One can bewail that, but not change it!"**

Such a development of character is possibly the easier, the more someone likes to calculate and the better he does it. Pressler has experienced it with himself. It has occurred to him, although he calls himself an "intensive nature lover," to talk about the "unfortunately only aesthetically noble beech," as if this beautiful tree species did not deserve in a hundred other ways the adjective noble, too. Namely, he saw "in every tree and stand a capital of timber and at the same time a smaller or larger timber factory, a factory, whose journeymen (root, trunk and leaves) have to earn their annual feeding (annual interest of timber plus the basic capital) by their annual labor (increase of value) and which, if they cannot do this anymore, have to give way to other, more productive workers."

But it is a strange thing with the factories. However appreciable their products often are, they themselves are unpleasant to us, for good reasons, and as a rule they are not treated with particular sensitivity by the owner. The continuous calculating and striving for higher profit may suffocate higher ambitions. If the owner of a factory is a character who is destined for the pleasure of beauty, he usually doesn't try to do justice to his better self **in connection with his profession;** in the best case he is going to look for a substitute within the fine arts, which mostly is going to allow only a few persons close to him to join the pleasure.

If, however, the theory of the highest net revenue wants to cleanse itself from the suspicion that it is not serious about the forest's beauty, then it has to admit, not merely occasionally (in fine print) that the "granting of individual pleasure" can be calculated as an income, but it has to state how to do that, and—this is the most important thing—then it really must calculate according to it. Here the popular saying fits: "Then no pursing of the lips is valid, it has to be whistled."

It is possible that one can end up with a right decision without calculating; but if one calculates and then omits an important quantity, then the result of the calculation has to be wrong under any circumstances.

It would be the most in the meaning of Neumeister to raise the rotation age by some years according to **expert's estimation.** The esteemed author of the *Forest Organization of the Future* raises "for safety" the rotation period calculated to be the most advantageous by the quite arbitrary time period of five years. So he could just as well raise the rotation by another five or 50 years, according to the circumstances, to increase the already favorable aesthetic pleasure. Another practicable way would be to **reduce the forest-friendly interest rate,** by assuming that the forest is going to raise a part of its income by granting aesthetic pleasure. It is well known that every reduction of the interest rate has the effect of raising the financial maturity.

So for example Poepel writes: "We do not intend to extend the theory of the highest net revenue management to the extent that a pleasure or an advantage like hunting would not be permitted; the principle of the highest net revenue does not need to suffer from that. A part of the interest rate or interest income lies in the costs of this pleasure itself."

Such a reduction of the interest rate would also remain a matter of emotion. Poepel himself does not even suggest a certain measure of aesthetic reduction of the interest rate, not to mention trying to justify it.

Anyway, it would be more in the sense of the strictly mathematical theory of the highest net revenue profit to study **the laws of the increase of beauty.** One should concern oneself with not only studying the teaching of the inner-forest **a** and **b** increase of beauty, but the outer-forest **c** and the principles of the half inner, half outer increase of beauty. Its relations, however, are complicated, because on one hand it takes place in the timber stands with increasing age, while at the same time, with the increased development of plantation, the treasures of unspoiled nature gain more value for us each day. R. Hartig seems to think of this way of calculating as particularly appropriate when he expresses himself like this: "But I would like to trust the adherents of the highest net revenue theory not to calculate the determination of the cutting age of a beautiful old stand and its percentage rate of value increase only from the sum of the volume increase, the quality increase and the increase due to rising prices, but to add quite a high percentage for **beauty increase,** especially at places which are easily accessible to the public."

In putting a value on the pleasure which the forest offers, one might start out from known quantities, for example the admission charges which are demanded from the visitors at art exhibitions or at concerts; only it would be difficult to estimate where one can find the higher pleasure, whether to find it in the stuffy hall or in the forest. As far as I am concerned, I would like to assume that with otherwise equal conditions, a forest managed on a 120-year rotation period possesses three times the attractive power of one which is subject to 60-year rotation periods. For the attractive power grows with the increasing beauty which is inherent in the mature forest, and it is in an inverse relation to the size of the cutting units.

Recently Hufnagel has expressed himself in this sense: "The

pleasure and satisfaction in the sight of old, tall stemmed timber is enough to let some owners stay with high rotation periods."

Certainly, the rotation period of the highest future property value would not be different from that of the theory of the highest net revenue, if one would be serious about introducing aesthetic values into the calculation. Whether one is an adherent of the highest net revenue theory or not, in any case the forester might want to keep his sense of ideals!

In contrast to the pessimism of Pfeil, the unforgettable State Head Forester von Hagen proved that this is possible even in a responsible high position, when I heard him speaking in a solemn occasion:

"We are priests, priests of the forest, dedicated to daily worship in a temple which has not been built by human hands, but in a temple which God the Lord has erected himself, and we faithfully keep up with this service as a priest of the forest; this is our honor and pleasure."

Chapter Six.

The Regeneration.

In the *Munich Illustrated Broadsheet* (No. 398), it can be seen how the barber has finally cut off the top of the nose of his properly soaped and rather scratched victim. But he knows how to help himself; he washes off everything quite well, puts on a bandage, puts on more bandages, and then he puts the mirror in front of him, with a face full of satisfaction and pride, as if to say: "Look, farmer, don't you look good now!" This example of the self-satisfied barber shows to us how large the inclination of humans is to judge their own performances particularly favorably. This is the source of a satisfaction which, however, cannot really be called an aesthetic one, but which can come up to the latter one essentially supporting. Out of this source—the **pleasure in the self-made things, in**

our own performance—stems a major part of our preference for the plantation system. How could it otherwise be explained that we indifferently chop off the beautiful advanced reproduction in order to plant some saplings in its place, although, with their mutilated roots, first they are not going to grow at all and probably not very well later, either, or similarly, that we do not prefer to let an unkempt hornbeam or something like that remain in the pine stand so that it will green the base of the balding pole stands by its production of seeds in the future, instead of experimenting with artificial understory.

Under more of an aesthetic point of view than a purely practical one, it will often be more important that the right **form of regeneration** is chosen—**artificial** or **natural regeneration.** The latter, if managed with knowledge, enriches the forest with many and interesting scenes by the diversity of appearances which the stands show **from the first preparatory cut up to the final clearcut. Plantations** managed in this way, **not looking all alike** and each one with different groups, are also pleasant sights right from the beginning. Beautiful natural stand mixtures are surest to succeed where one lets nature itself establish or at least influence the mixture. If this natural establishment of seedlings does not succeed, then one should not hesitate with artificial support in order not to let conditions develop, like those that drove Wildungen's old forester rightly desperate.

Regeneration below sheltering trees, by the way, offers opportunities to form good trunks for old standards and to get them accustomed gradually to free position, so that they will soon be an adornment of the stand, supplied with a regular crown and being without water-sprouts, like the pictures V and VII show. Due to all these advantages, natural regeneration deserves to be preferred under the aesthetic point of view everywhere where the conditions make its success nearly certain; in other cases a well-groomed **sowing or planting** is more appropriate. However, it must be required that it establishes itself and grows without gaps and disease right from the beginning. If it unites the impression of rapid juvenile sprouting power with neatness and order, so even a planting of one-year-old pines (although maybe already, with smashed roots, carrying the germ for premature death in them) can outstrip the thin-needled, often damaged, seemingly incomplete groups of a natural regeneration for quite a while.

The practice of combined cropping [growing trees and field crops on the same area] shows the strongest juvenile sprouting power in comparison to many other forms of plantation. In this respect it deserves the appreciation also of the aesthetician; but the advantage faces greater disadvantages, namely the extermination of the wild flora of shrubs and perennials, and the transparency of the stands planted in the direction of the plowing furrows.

One might ease these disadvantages, if shade tolerant deciduous trees (hornbeam, hazel etc.) in a moderate number were added to the main tree species for accompaniment from the beginning.

Whether broadcast sowing or strip sowing, planting in rows or planting in regular pattern is to be chosen, depends on many circumstances. In most places, an appropriate variety will be the most welcome. Anyway, it would be wrong if one wanted to plant **all** stands in a district in a triangular pattern and thereby cause the effect that every place appeared transparent far into the interior from each side immediately after the first thinning. This would cause the forest to lose much of its magic for us, and wildlife would not feel comfortable either. But every now and then, one may do something else; for example, at a property line bordering a mismanaged forest, a planting with balled plants, neatly set in a pattern, might give a perfect example for the good order loved on this side of the line. Besides the advantage of cleanness, the regular patterns, and especially the wide-spacing, may be praised because, due to their openness, they give an imposing impression early. To this fact we owe the knowledge (the only one I know of) that Georg Ludwig Hartig also had a feeling for forest aesthetics. He writes about a 70-year-old stand of spruce, "which had been planted and grown in nearly rod-wide, perfectly fitting rows, totally symmetrical":

VII.

Postel. Horstweise Eichen-Vorverjüngun mit Kiefern-Ueberhalt.

"Besides the great advantage which this planting offers to the owner, it also makes the most pleasant impression on every nature lover. I have to admit that I have been utterly pleasantly surprised by the outmost regular stand of such large diameter and tall trees which are grown straight upwards and have no branches to a height of 70 to 80 ft, but at the same time have a completely closed canopy and form a magnificent bower."

If one, however, as I said above, as a rule can handle it the way one likes to, I have to stress: **Small areas with irregular boundaries within forest stands with irregular patterns are by no means allowed to be planted in regular patterns.** This is valid, for example, for gaps caused by windthrow, for grouped advanced regeneration and grouped undergrowth. After all, the young plants are supposed to grow together with the surrounding stands into a harmonious entity as soon as possible. In order to achieve this in the fastest and safest way, one might consider a group of naturally established young trees as an example, and should put into the center some taller individuals of an intolerant tree species, at the edges smaller plants of a shade-tolerant tree species, and may make the latter individuals gradually disappear into the neighboring stand.

Whoever does not want to do without straight lines, even under these conditions, may draw his planting line parallel to the road in order that one might not be able to look into the planting rows from the road. The most transparent ones are straight-lined plantings of saplings. Particularly for those, the following rule of Prince Pueckler is valid: "Concerning the thorn bushes, I always have in my mind's eye the instruction of Herr Repton, the excellent garden artist, rarely to plant any tree without giving it a thorn as a protecting company. Even if this is not to be taken literally, nothing can be more useful indeed as a protection as well as an adornment of the plantings."

Hampel has pointed out recently that regular plantings may be suitable here and a mistake elsewhere when he wrote: "The people after all need the forest for their well-being. For areas like these, or those which lie within sight of country estate homes, castles, etc, one should not use regular plantings, but irregular ones; for the former cause a monotonous effect and provoke one to continuously count the rows, they disturb the mind instead of offering a piece of rest for the eye and a mental pleasure. In a great system, regularity has a quite different effect; it shows a beautiful order and is never displeasing, otherwise the fields with their furrows and crop rows would have to be unsightly as well; but where there are vistas or where they can be opened up, one should tend to bring into an irregular stand particularly the valley sides close to the vistas or opposite to them."

Where the consideration of intervening agricultural cultivation does not force one to ongoing straight lines, one may avoid **excessive** transparency, for example according to the patterns of Figures 33 to 35.

It would become really dangerous with rows of larches planted at a distance of 10 to 15m as recommended by Boden; it would be even worse with a sprinkling of larches in an 8 to 10 square meter pattern, which he also approved. By this sprinkling, rows of fast-growing, immature larches would evolve into every direction, and they would cut through the slopes in a checkerboard pattern visible from miles away.

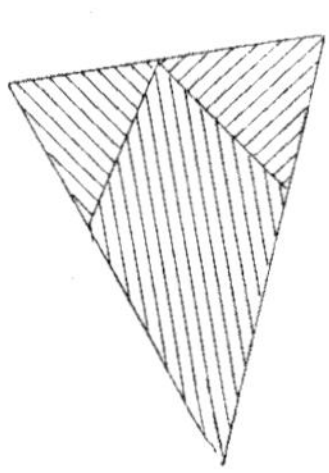

Fig. 33

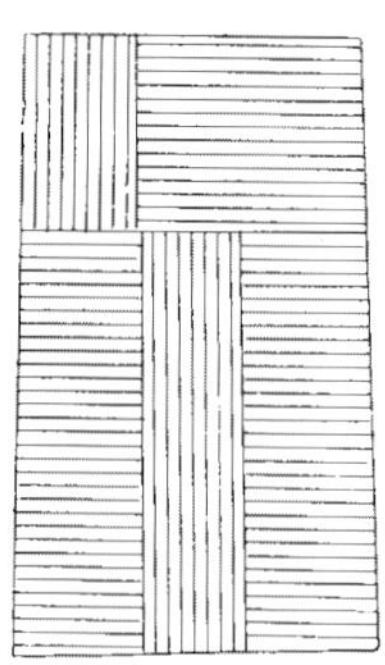

Fig. 34

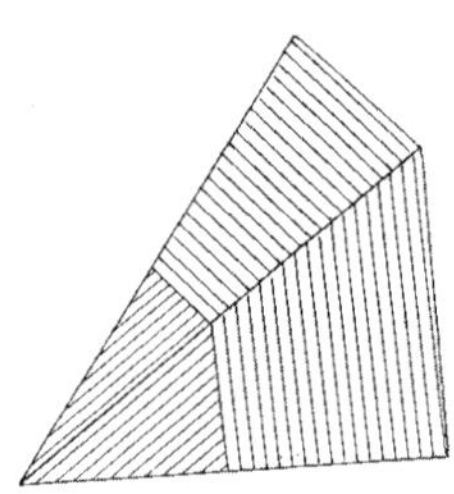

Fig. 35

In respect to aesthetics, **mound planting** deserves to be preferred to **hole planting.** The former satisfies insofar that the culture appears more impressive while it is being created. If the mounds are large, then an even greater advantage is gained in the future, because the beauty of older trees is dependent very much on the visibility of the strong roots. The higher one plants and the more one can see of the strong roots, the more stable the tree seems to be rooted in the ground. "An old tree which comes up from an even surface loses half its effect." Gilpin realized that long ago; he would not have agreed with the area around the lion's monument in the Berlin Zoo, where the soil between the trunks is raised each year by fertilizing with compost.

For the planting procedures and also for possible corrections, **germination beds and nurseries** are to be maintained, but it is the question whether permanent **fields or temporary fields** are to be preferred. One will probably like to see the latter if they bring the only bit of variety into the monotony of vast, remote plantation areas, but one should put them a little bit beside well-used roads wherever possible, in order not to make the weaker tree growth on the worn out soil of the field area unpleasantly visible during every walk. One can avoid completely the unfavorable sight of worn out field areas if one uses them not more than twice, and if just in the first year, one distributes irregularly some deciduous saplings on the area. These usually develop very lushly on the plowed soil during the care of the field.

Permanent fields at the right place may be an adornment for every district. Nicely fenced with hedge or fence, streaked with borders on which some rarer tree species, even flowers (snowdrop, cowslip, liverwort, aquilegia and whatever else gives a nice border without much care) can find a place, fields like these can become small **forest gardens** and thereby the main adornment of the forest district.

As much as one enjoys the success of one's own industriousness in the appearance of a thrifty young plantation, one wishes to forget the human efforts and worries in older stands. One would like best to consider them as a free gift from Mother Nature. This impression of a mature wood stand is destroyed the moment economic management intervenes. Therefore **row thinnings** have to be condemned, and the **gaps cut in regular forms** in order to establish groups of advanced regeneration are very disadvantageous for the aesthetic effect of the mature forest stand. The group-wise advanced regeneration therefore should be started always below shelter trees, which are to be thinned out gradually, and the release should happen only quite according to the needs of the regenerated tree species.

Picture VII shows a group of oaks treated in this way.

In establishing a stand one should avoid any permanent **reshaping of the soil surface** in order to make the artificial creation unnoticeable in the future. The combined cropping should not plow up **beds,** and deep **forest plowing furrows** are to be avoided. Not unsightly because they are invisible, but therefore the more serious, are the smaller and larger **holes** which unfortunately can be found all too often between planting mounds. Due to lack of prepared plantation soil, the forest workers take the soil for the mounds out of the plantation area, but they neglect to make these holes safe by flattening the edges. The pitfalls are dangerous to humans and animals. Young forest birds are going to die in them as well as the useful ground beetles; deer will avoid such deceitful areas, and to the hunter himself, a stand is spoiled where one runs the risk of spraining one's foot with every step.

On the other hand, the eagerness to make everything neat goes too far if one carries out **stump removal.** Once I listened how the head forester called to the forester: "But why did you not clear here?" The forester answered: "Herr Head Forester had given the instruction to cut in order that we get finished earlier and to avoid having a trunk fall into the area with young plants." Then the head forester said: "Quite right, that is what I have instructed, but you ought to have known that you needed to make an exception right here at the country road. Now this looks too messy and wild."

According to my opinion the forester was right. Old stumps do not spoil the forest; quite the contrary, they are rather picturesque in their rough forms; dressed with lush green moss, with colored mushrooms, they are an adornment of the foreground, which the painters know to appreciate. Particularly thick stumps are also valuable in another respect. They show what the site is able to produce and increase the pleasure in the young woods in the hope well founded that the small plants, though still weak, will grow up to be forest giants sometime in the future.

Therefore one might want to let not all, that would be wastefulness, but some of the stumps remain in the areas that were cut to be regenerated.

The final picture of this chapter may be a monument for the once magnificent oaks of Spitzberg [pointed mountain] in Katholisch-Hammer, whose last remnants it presents.

Fig. 36

Chapter Seven.

Taking Care of the Stand.

Weeding, thinning, and pruning are needed to guarantee the vigor of the successfully established timber stand. Unfortunately, these actions do not leave a pleasant scene immediately after their completion, except to the trained eye of the forester, who does not look at the things as they are at the moment, but as they will be in the future due to the present destructive operations. However, ordinary people notice that a pine has suffered from an abrading birch, only after the lopsided pine has been freed from its torturer; and also the recently pruned oak with the shimmering black painted scars of the branches is not for everybody's taste. Likewise, recently thinned stands will always look somehow unnatural. **Because one can not avoid this and similar operations, one should concentrate them at one place,** in order to allow the rest of the area to enjoy peace.

I have successfully tested the workability of these practices for 25 years in different districts. I divided my own forest of 665 ha into four districts, and each year one of them is managed, i.e., cuttings, thinnings, prunings, road maintenance, etc., take place each year only in one and the same district, while three parts of the forest enjoy complete rest and quiet. Particular circumstances justify sometimes variations of the institution, which is proven and tested in the whole.

So much in general. In regard to the three management operations, combined in this chapter, as individuals, one should not demand too much rigor concerning the **cleanings;** quite the contrary, one should require the manager to reduce the harmless shrub species—willow, elder, roses, etc.—only by the amount which is really necessary for the well-being of the plantation. **If then the thinning comes early enough, quite a few shrubs may survive as an useful and decorative groundcover.** The same consideration should be given also to tree species which are able to grow up in the closed stand until the second or third thinning. Mountain Ash and Goat Willow are preferred by me due to their beautiful blossoms and their fruits. I also save the aspens as much as possible up to usable diameter, and for severe winters, in order to be able to cut them as a natural wildlife food. With some attention, these tree species, which get rarer and rarer, are very well worth the consideration given to them without becoming awkward. Similar things, if not the same, will be permissible at many places, but everywhere one might want to require as a guideline the ideas mentioned above about mixed stands and stands decorated by individual embedding for the rules of stand management, as well as for the establishment of stands.

The traditional **thinning management** ("the suppressed has to be cut, the less vigorous may be cut") had the mistake of letting the stands become transparent very soon, which was unsightly as well as undesirable from the hunting aspects, and had as well some forestry disadvantages with it.

In applying the **Postel thinning method,** this failure (that the stands become bald at the bottom) can be postponed for beeches and firs for a long time, for pines and oaks at least for some years.

My method, in short, is the following: **I start as early as possible with the first thinning, but restrict myself to give space to the crowns of the dominant little trunks (Class I) by cutting out the crowns of the less vigorous (Class II); the suppressed ones (Class III) are saved.**

It is understood that under ordinary circumstances, not all of the Class II stems are cut in **one** year, but with an appropriate—about three to five year—return of the axe.

Soon a very considerable difference is going to form between stems in Classes I and III. The latter ones, eking out a modest existence nearly without noticeable growth, will stay unmolested in future; the former will be thinned out in a four-year cycle, moderately in the beginning, later heavily.

The picture VIII shows a beech stand, thinned in this way 25 years ago and four times since then.

The theory with me is a child of the need. In a beech thicket of saplings near the well-visited Johanna's Height, there are tables and benches on which in the warm summertime the urban citizen likes to enjoy his coffee or beer or whatever. If it were not managed, the beeches soon would have become transparent, and then the site would have remained neither so pretty nor so protected; it would

VIII.
Postel. Posteler Durchforstung in Buchen.

have remained transparent at least for two generations, until one would have been able to help a la Seebach. There I have done the first test, and this with so much favorable success, that even today, 25 years after the first intervention, a part of the suppressed little trunks has stayed alive as a desired understory.

At the occasion of the meeting of the German Forest Organization at Wiesbaden, Ney said:

"Now our people are trained to look first whether the future stems need any help, and it is prohibited by death penalty to cut out something at the bottom [the understory], as long as there is still something left in the top [overstory] to be taken away."

After such decisive support I may at this point refrain from defending more in detail from my side the saving of the suppressed little trunks.

The Borggreve selection thinning does not offer pleasing forest scenes in the first years after the cutting. The action, returning every 10 years and removing 10 to 20 percent of the wood by volume, taking out mainly the larger diameters, makes the stand appear thin and much younger than before the cutting, while other methods which take out less vigorous individuals seemingly and sometimes actually raise the average age of the stands.

Some years after the cutting, however, when the crowns have nearly closed, a selected, thinned out stand often looks very good. It gains an elegant look, because mainly the straightest and least-branched trunks are left. Under the forest aesthetic point of view I therefore have to wish that the interventions in the closed stand are practiced in a less obtrusive way, in the sense that instead of cutting 10 to 20 percent every 10 years, 5 to 10 percent are cut every 5 years. A considerable raising of the rotation period has to accompany the introduction of this method, as Borggreve explicitly stresses; otherwise it will cause severe disadvantage also in respect to aesthetics, while it produces very beautiful forest scenes in fulfilling the conditions above, as the experiences described here teach. However, I pay attention to **preparing the stands for selection thinning by**

beginning early and repeating often in such a way that also the less vigorous individuals have good to rather good crowns, which develop quickly into good forms after the release cutting anyway. Picture VIII shows a pine stand in the background which has been selection-thinned repeatedly.

The marking of the thinnings is an exhausting activity, if large areas are to be done quickly. The neck aches, the eyes get tired, the whole body gets weary. But when the work is done, the forester does not easily leave the stand without looking back with satisfaction on the result. Visible from the distance, all at about the same trunk height and placed on the same side, the fresh blazes shine towards him. The impression of clearness and order fills him with deep aesthetic satisfaction, which is even raised by the connection by ideas. The trained eye views the stand foreseeingly in that improved condition as it will look after three years, when the freed crowns will have widened. We also hope to thin more often and to produce something really perfect. Not much is lacking, and we will speak like Faust:

> *"In anticipation of such great luck*
> *I already enjoy the greatest moment."*

Such life in the distant future is a reward for our efforts; but we should not imagine that the layperson is able to understand our feelings—this one does not think of Borggreve's wolf trees nor of Kraft's stem classes; he regrets that it is to be cut, and each blaze threatens a dryad with disaster.

We need to take into consideration these feelings; therefore the thinning axe has to be handled with caution in a way that the layman does not notice the preparation for the cutting when he walks by, along public roads and anywhere else the public is present. Of course, the statement is valid even more for marking trees for cutting; for the larger the tree, the greater the compassion which is given to it. In cases like that, one should blaze the trunks only on the side not visible from the road, or only into the bark, if one does not want to delay the marking until right before the cutting.

If the stranger passes a forest where nowhere does one tree grow together with another into a poor shape, where no forward branch rubs off the bark of the neighbor tree up to the cambium, where spiral-grained trunks, forked trees, etc., cannot be found at all or only rarely, he soon notices that here an especially caring hand is at work. Such an awareness touches pleasantly not only foresters, but laymen nearly even more, because they pay less critical attention to details but abandon themselves more uninhibitedly to the impression as a whole.

The most efficient activity in this direction is prunings, if they are handled with caution. Spruces and pines, grown in open spacing and neatly branched, look very good after the healing of the wounds, nearly elegant; **(the true elegance—I read it in a fashion journal—namely requires faultless performance with the best material in avoiding anything superfluous).** Even greater is the attraction of pruned oaks in free position or of beeches grown with only slight pressure from the side; even more than trunks equally clean-boled in dense closed canopy these appear as if they had chosen the elegant shape by their own will, not forced by the struggle for existence.

Clumsy pruning, of course, will provoke just the opposite impression, and, even with perfect performance, **the care of the stand has its limitation where, if it is exceeded, it will damage the area's beauty. Namely at places where nature unfolds itself especially magnificently, one likes to see its forces working freely as much as possible.** Therefore in romantic mountain sites one should not be too quick with the thinning axe, not to mention the cross-cut saw. There it does no damage at all if every now and then even a windthrown trunk is left lying, providing a welcome home for ferns and mosses and a rooting place for the germination of a new spruce generation.

In a pleasant way it is possible to serve at the same time the forestry point of view (the desired elegance) and the taste of the naive nature lover, where there is a particular **rich groundcover;** for this is going to form itself the more diverse, the more free and lovely, the more emphatically the closing is loosened up by axe and saw in the stands. Seemingly owing its existence to the willful play of coincidence, it presents itself as a source of pleasure, which the overly cultivated forest of the plain and the hilly regions otherwise might easily lack.

Up to now I have carefully avoided going into technical details. Assuming forestry-educated readers, I have presupposed the craftmanship rules as known. Here I make an exception with pruning, for the pruning action is still too much at sixes and sevens with us, and prunings inappropriately done misshape the trees severely in aesthetic respect.

Some basic rules are generally known: One should prune only during the time without sap flow; one is not allowed to leave stubs standing there, but one has to lead the straight cut very close to the trunk in a way that the trunk is not going to tear. The wound has to be varnished over with hard coal tar. Branches of more than 7 cm thickness are not to be cut off at all. All this sounds very nice, but in reality one too often ends up in the situation where one is forced to cut stronger branches, and this is then alarming.

Therefore, it is important, first of all, to prevent the establishment of strong branches. Like the Frenchman says: *gouverneur c'est prevoir* [to govern is to anticipate], so also here the main art in pruning lies in the right foresight, in the correct valuation of the development, which the crown of the tree is going to take.

Before the tree climber starts his work, he should take a good look at the crown and decide which shoot would be particularly suitable to serve as a continuation of the trunk. Then he should watch whether any other shoots gain a lead before this one, or tend to an excessive strong development. These are mostly the branches at a steep angle, the so-called fork, and other branches which are a little bit steeper set than the rest of them. These have to be fought first of all, in order not to let them get even stronger, so that their future removal would cause overlarge wounds. So it is completely wrong, to cut always the lowest branches first. If there are strong branches in the upper part of the crown, then the removal of the weaker branches below them will cause an increased streaming of sap upwards, and the tree actually is forced to nourish even more the already over-developed branches, and often the formation of dangerous lateral tops is caused or at least favored by such wrong actions. If there are more of the dangerous branches than one should remove at the same time, or if they are about equally as strong as the leading shoot to be favored, then one should hinder first the growth of all of them by cutting the tips. It is the right way, if one grows trunks which are trimmed only by very many but weak branches.

These rules are valid for all the tree species.

Picture IX shows an oak successfully groomed in this way.

On artificial roads which cut through the forest but are not subordinate to the forest administration, unfortunately one can very often see road trees which have been mistreated according to the instruction by Parisius, which is meant to be for fruit trees, in deliberately cutting away the center shoot and preventing its substitution. This can be seen even within the training forests near Eberwalde! The forester should not close his eyes to the sight of this tree abuse, but lead to an improvement by education and example. In two Silesian and one Markian districts, I myself found it easy to change this, in showing the Landraete [administrative heads of the Landkreis] how to do it better. In several cases the **completion of the missing leaders** succeeded very soon, after I had recommended to remove out of the group of forked branches each two by a diagonal cut. At the tarred wound's edge then strong water sprouts developed, one of which could be used for the formation of a new leader. I have practiced this method with young oaks, elms, maples and apple trees. Mutilated older trees should be cut down, the sooner the better, and replaced by new planting.

I am going to close the reflections dedicated to prunings with the remark that one should grow the stands in a way that there are as few trees as possible that will require pruning. They should be kept well-closed in their youth, until they have pruned themselves naturally. Small trunks which grow irregularly should be rejected already in the nursery, and in the stand they should be chopped off unless the cutting will cause severe gaps.

Chapter Eight.

The Secondary Utilizations.

The forest litter utilization is very disadvantageous, practically as well as aesthetically. The interested reader might want to leaf back to Page 48 and realize in which way nature presents "blended colors." This happens in a particularly magnificent way in forming the soil covering. Residues of deciduous leaves and needles, mixed with dead branches, broken through by mosses and other delicate plants, form an inimitable carpet. I write "inimitable" quite intentionally, and whoever has tried the imitation is going to agree with me.

I have tried repeatedly to combine material brought home out of the forest into natural groupings for foreground studies, but I never succeeded in doing that.

This failure becomes understandable if one knows how carefully nature weaves its carpet. Ebermeyer has transmitted a graceful description of the fall of the leaves to nature lovers as following:

"Without being moved by a breeze, one leaf after another removes itself softly and silently from the branch, out of the colored tops, golden with the light of the fall sun, and falls, or better floats and dances in beautiful circles down to the earth. And every tree species has its characteristic dance of the leaves. The heart-shaped leaves of the linden, which go down to earth so early, swing themselves down differently from the lobate leaves of the maple or the hand-shaped fans of the horse-chestnuts. With all of them, the course describes a graceful spiral, but the turnings of each species have their own form, according to the laws of equilibrium which happens between stem

IX.
Postel. Eichen am Militscher Weg.

and area of the leaf. At the same time, even from one and the same tree, no leaf falls down exactly the same way as its companions; the larger one runs faster through its last move; a leaf burdened with hoarfrost comes down to earth conspicuously faster; a third one falls upon a branch, rests there for a while and then moves down to the ground in the company of some fellows, which it has inspired to the move by a soft touch. One might watch for hours and would perceive always new beautiful movements of falling."

Something that has gone to rest so gracefully should also be left alone there; otherwise one deprives the forest of much of its beauty.

Multiple connections of ideas unite in order to express this wish, for Homer sings with cause:

> *"Like leaves in the forest, so are the generations of humans.*
> *Leaves are blown to the earth by the wind now; others are grown*
> *By the budding forest again, when spring newly lives up;*
> *Like the human generations, this grows and that disappears."*

But we do not need to go back to the classic age. Whoever has the Kommersbuch closer at hand than the classic writers will find already in the second edition the appropriate appreciation of the forest's litter. There Theodor Koerner *("The Five Oaks of Dallwitz")* addresses German oaks like this:

> *". . . And when in fall your leaves will fall,*
> *They are a precious matter to you even dead;*
> *For decaying, your children will be*
> *The foundation of your next spring splendor.*
>
> *Beautiful picture of old German loyalty,*
> *Like it has seen better times,*

Where in joyful keen consecration to death
Citizens have established their states. —

Oh, what does it help that I renew the pain?
All are familiar with this pain!
German people, you most magnificent before all,
Your oaks stand, you have fallen!"

The beautiful allegory has not betrayed. Soon after the verses had been composed (1811) the poet fell for his country, but "he is a precious matter to it even dead!" We honor him together with the thousands of his fighting companions out of old and new days as the "founder of our present spring splendor."

Not only unpoetic, but unaesthetic in every respect is the forest with litter utilization. This use, damaging the site, reduces the height growth of the trees and therefore not only the revenue, but the forest's beauty at the same time. Many forests of small estates, devastated by litter utilization, show this in an abhorrent way.

However, it may be quite advisable under circumstances to rake litter on the forest roads, because the roads become more walkable by doing this.

Grazing, utilization of berries and mushrooms, of windfall wood and dry fallen wood altogether have in common that their practice disturbs the forest.

It harms the idea of forest solitude if one can suspect behind every bush a being whose appearance has little in common with the ancient ideas of nymphs and dryads. In respect to grazing and windfall wood and dry fallen wood, more or less the same is valid that was considered for saving the forest's litter. On the other hand, it cannot be denied that it would be irresponsible, under moral points of view, if one would deprive the mostly poor population of the forest areas of an old traditional, often irreplaceable source of income, as it also raises a conflict to see appreciable natural products, for example bilberries, rot without utilization.

I think the right solution has been found where the forest owner **gives the intended utilizations to people who work in the forest as well as to such people who do not have another appropriate source of income at the moment.** Good supervision has to take care that **there will not be disturbance every day and at every place.**

It is particularly disgusting, if old women crack [collect firewood] in the stands for hours, in order to pant home then, a picture of misery, under high burdens of half-decayed twigs, as a bad omen for the hunter moving out to the woods. Therefore quite a lot of cubic meters of cordwood should be delivered right to the front door of really needy persons, but thereafter nobody should be allowed to go cracking.

I write this in a region rich with woods. Otherwise Wilbrand is right with the remark that one is not allowed to let anything rot and that it is disturbing to the aesthetic satisfaction if dry twigs decay which might help the population's need for firewood.

In the Bohemian domain of Czleb I found this attitude realized in the most perfect way; the population in need of wood sawed off every dry little branch up to a considerable height of the trunk, close to the stem. If I remember right, the forest administration provided the necessary pruning saws, and the people were glad to be allowed to carry home the wood for their labor. This, however, looked very neat, nearly too neat.

Cattle in the forest do not go very well with the intensive management of modern times. Goats, horses and sheep will mostly be a horror to the forester's eye at places where trees stand or are supposed to stand, and cows will often be a horror. Only in very large forests, herds, together with the peal of bells and shepherd boys, will be welcomed nearly joyfully as a brightening element. Here it cannot be generalized. It will depend upon the size of the district, the type of management, the abundance of wildlife, the power of the soil, even on the whole character of the region in which limits to the grazing utilization may be permitted.

Forest grazing has found a warm admirer, standing in the

"field of emotionally warm contemplation of nature, uninfluenced by rational reflections" in the person of Tessmann. Reporting on the forest in Scandinavia, he writes that "the domestic animals present a friendly as well as necessary element to the northern mountain natural area, nearly overwhelming by its size. There the herd of spotted goats brings life onto the slopes grown with brush. The raffia whistle of the shepherd boys, the bright bells of the cows welcome us as a homely, reconciliating signal after hiking for hours in the monstrous districts as silent as death. Every traveler knows what a melancholy weighing sound is lying in fall above these forest areas and rock districts, when humans and herds have moved back into the valley, when one passes the closed summer cabins and everything becomes more and more lonely. No friendly breeze breathes at us, no familiar sound touches our ear, only the croaking of the hungry bird of prey mixes with the monotonous sound of the roaring of the cold ice water! Certainly, **with the banishment of the herds out of the gloomy forest districts of the North we would again lose quite a lot of mountain poetry."**

Similar will be true for vast connected districts of our country; in smaller districts one will still like to see one or two cows, grazing from a rope in the firebreaks (of course, these are the cows of the forester or the forest guardian). Even these are not good where great order and neatness have to exist, nor on steep mountain slopes.

In mountain regions, of course, the grazing utilization is particularly important in the economic and aesthetic respects, this in a good sense as well as a bad sense. Where on steep slopes with soil types which erode easily, the cattle cut their tracks deeply into the soil, there the trained eye already sees the future disaster: sooner or later there will be a thunderstorm, and as a consequence the mountain slope will be bare of soil, and the valley will be buried under fragments of rock. Who does not possess the happy character, as at one time Rueckert's *Man in Syria,* will not be able to look even for a while at the magnificence of the Rigi [a Swiss mountain] without being disturbed in his aesthetic pleasure by contemplations of this kind.

Quite different is the cause with the ground of the valley and on softly-inclined high mountain pastures whose lush sward is more stabilized than loosened by the step of the cows.

Such pastures, not to mention the accompanying herder's cabin, and various pleasant associations of ideas which are caused by their sight, belong to the most beautiful that the forest holds. As seen from the distance, they moderate by their light green the impression of mountain nature, which is easily too serious; coming closer, one finds the attractions of the park and the garden united in them.

Zeidel-grazing, that is beekeeping in the forest, was, in former times, a reservation of the landowner. Nowadays the forest owner is going to leave this line of business to his employees. The beekeeping is of aesthetic importance in that it encourages the embedding of the beautiful honey producing trees (from hazel and maple up to *Robinia* and linden) into the stands.

If I wanted to draw **hunting** into the circle of discussion, this would be the place to think of it; however, I have too much respect for noble hunting than to think that I could be allowed to treat it only in passing, in one single chapter as hunting **utilization,** following the secondary forest utilizations.

If not an independent book, then at least **its own part in *Forest Aesthetics* can be claimed by hunting aesthetics,** but I lack the time, the necessary knowledge, literature and the leisure to write it.

Chapter Nine.

Meadows, Water and Fields; Forest Edges, Hedges and Fences.

If a forest includes meadows, water and fields, then these have to be related to the larger entity as much as possible, in order to create a harmonious impression.

In many cases the forester has the opportunity to form the outline in a beautiful shape. It is his task to judge whether he should here add some acres of alder swamp to the meadow, whether this little soil mound, which burns out all the time, should be used in order to fill up sinks nearby, or if it is better given to the cultivation of trees, and more of the like. In short, only rarely will it be found exactly marked to the meter on the terrain: up to here it is advantageous to establish a meadow, and not further on.

These so-called transition zones have to be used in order to **gain shapes and outlines as beautiful as possible.**

The guiding points of view herein are the following:

1. **One should never see the ends of the area,** so it will occupy the imagination more and it will appear larger than it is.

2. **Indentations of the edge have to be formed so the morning sun as well as the evening sun will be able to shine broadly upon the grass area and, if possible, on particularly beautiful trees,** while the majority is lying in the shade. **The projecting and set-back angles need to be clearly visible;** due to that, they also will actually cause pleasure. What has to be avoided the most are lines winding **in the same way** for long distances (corkscrew lines).

3. Care must be taken that **particularly beautiful trees with many branches either form the edge themselves, or that they step forward individually in front of a closed edge** and gain a beautiful background in it. If there is a lack of stand trees which simply might be left standing, so that one still has to plant and grow these, one should choose as a site best suitable for that the neighborhood of a block of stone, of a spring or the like. **Into the planting hole some bush (hawthorne, cherry, elder, etc.) should be set together with the tree** in order to give the planting from the beginning the appearance of natural formation. But here I would like to remind the reader once again that the most beautiful single trees are not grown in free position from youth on, but in closed groups, except for spruces and firs.

4. **Entire portions of the forest stands are to be maintained in the form of an island at suitable places, respectively to be established** where the soil invites one to do so. Groups like these mostly will not be allowed to lie in the middle of the meadow, but

X.
Fürstenwald bei Ohlau. Di Weinert-Wiesen.

they will have to present themselves as **runners of a larger part of woods,** like islands in the sea can be understood as attachments of mainland massifs. The meadows of the Silesian Oder administrations correspond nearly without exception with these demands, quite without the intention of the Forest Administrations. Picture X at the beginning of this chapter shows an example of that. I would like to draw particular attention to how nice it looks that the meadow area shows through below the separating stands.

One should pay careful attention not to spoil the natural impression of the design by unsightly roads and ditches. Even tame pastures often can be maintained without disturbingly visible ditches. In this respect, ideal is the **"branching of the water courses,"** as it is practiced on a large scale in the Forest Inspection Stettin-Torgelow. There, former bog areas have been set up for irrigation by surrounding them with a circular ditch, which nestles against the contours. The creek running through the swamp areas is dammed where it intersects the circular ditch, and the irrigation then happens by itself from the edges up to the old natural path of the ditch.

By the way, I should not forget to point out **that for improved pastures of larger extension, also straight-lined encirclement** by elevated woods roads creates a good impression, along which slim-grown trees find a suitable place. There everything is so neat and clear, the straight irrigation ditches fit so well into the straight borderlines, the surveying of the individual lots for the lease of grass is easy, the harmful shading can be reduced down to such a small amount, the artificial form fits so well to the concept of improved pasture, that one can only look at something like this with pleasure; indeed, everybody enjoys several of these creations in this region.

If the direction of the ditches is not dictated by the terrain, then one should advisably avoid setting up main ditches in a way that they are seen in their longitudinal direction. Otherwise they will appear too conspicuous and ruin the homogenous impression of the pasture. For the same reason **spoil of ditches** must be removed as soon as possible from the edge of the ditch.

XI.
Nesigoder Thiergarten. Ufer der Luge.

Above the irrigated meadows, **collection ponds** will, as a rule, do good service. Regarding the best shape for these, the same is valid that has been recommended for the pastures, but the outline of the water surfaces will mostly be predetermined by the terrain, and not much could be done to change it. However, one will be more inclined to set up a **dam** where a beautiful shoreline will be formed than where just a sack-shaped end of a small valley would be dammed up.

If the pond is formed partly or totally by **excavation,** then shaping of the shoreline has to be done according to good examples. There are two types of **shorelines** to be distinguished:

Waters which are freely exposed to the wind possess shores running in long-stretched lines, flattened by the lapping of the waves; smaller and more sheltered waters get their much more irregular forms by settled plants and by wildlife.

While alders, bushes and even grasses protect projecting corners, wildlife goes down to the water between them and flattens the edges of the bays, enlarging them. Where cattle grazing takes place, this type of shore-formation can be studied even better.

Picture XI shows a reservoir whose shoreline is maintained by alder roots.

The **dam** itself should receive a look as natural as possible by wide extended slopes and by avoiding straight lines.

Wide dams, curved against the flow of the water, also have the advantage of greater stability.

The dammed water usually is lead away in a so-called monk; the **closure of the monk,** however, has the disadvantage that one can only hear the falling of the water, but not see it.

For small ponds this failure can be avoided by a locally **tested form of damming,** which also has the advantage that it is cheap, can be closed very tightly, is very well protected against intruders, and does not need repairs. The construction is done by laying a clay pipe out from the deepest point of the pond down to the drainage ditch. The upper end of the pipe receives a diagonal-cut surface which will be simply covered by a board. This board is supplied with a hook so that it can be opened by a curved stick.

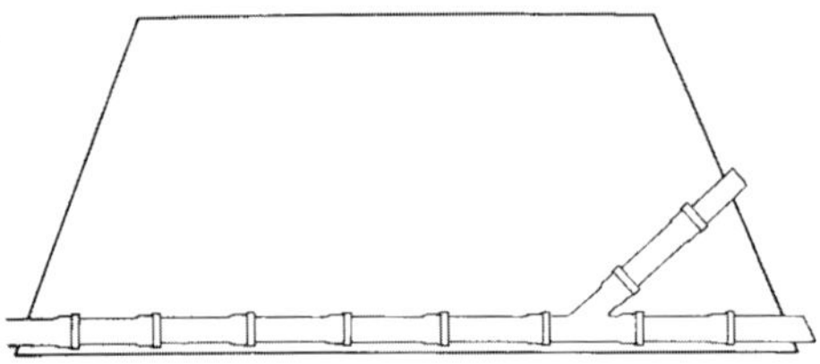

Fig. 37

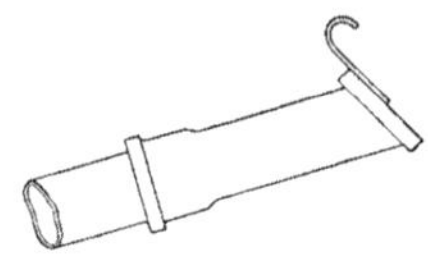

Fig. 38

Very deep ponds need an intermediate outlet of the same kind, like the figure above shows.

The water overflowing after the damming should be led not over the dam, in order not to threaten the dam, but into stable soil beside it at a place where the new water-course can flow nicely visible and without jeopardizing anything.

As important as the layout is the structure of the forest edge, enclosing pastures, fields and waters; but there is much disagreement about the **forest margin.** Some people demand branches hanging down low, forming a dense wall, while others want to show beautiful trunks. Some want long, wall-like forest edges to make a magnificent impression, while landscape gardeners say to design the forest edge interestingly by planting groups outside the edge of the main forest. Which of these recommendations should be followed depends on the place and the situation.

In front of me are pictures from the Holsteinische Schweiz [Holsteinian Switzerland], of the small Uglei Lake and Stendorfer Lake, whose shorelines are decorated with the green of beeches, hanging down low, and from the same region a picture of the Schuetten-Pond. In this picture, birches have become established individually and in groups in front of the beeches, and the picture is brightened up more cheerfully by this. In the Forest Administration of Donnerswalde, I admired one of the most beautiful forest edges: selection forest type spruce of uneven age, towered over by high-crowned pines, distributed in groups, which I saw shining in a magnificent way in the evening sun.

A pasture edge treated like a selection forest from the Forest District of Katholisch-Hammer is shown in Figure 39.

Fig. 39

Large trees form a main attraction of the forest edge, as for example the Susanneneiche (Picture II), in towering like a dome above the forest with their mighty crowns, interrupting the horizon line if seen from the outside, and forming an effective foreground if seen from the inside. The two beeches from the western edge of the Muellerheege, which are presented in Fig. 40, are visited often in order to admire the sunset as seen from the forest between their gnarled trunks.

It has already been mentioned above that it is an advantage of the composite forest [coppice forest with standards] that it forms varied forest edges. The diversity is nearly infinite, and in the individual case the decisive judgement has to be left to good taste.

Fig. 40

The things said here are valid quite generally for forest edges. I would like to dedicate a particularly short contemplation to **plantings near water.** It would be possible to write a brilliant aesthetic book of its own about the **dressing of the shores** of lakes and ponds, of rivers and creeks, but in the narrow frame of forest aesthetics, I would like only to add a few remarks on this subject.

Who would be so kind as to remember the things said of the aspen in PART ONE, like the pleasure we receive from their whispering in the wind, and the special feeling we enjoy when they rest for a while, but quite exceptionally, one might consider as similar but even more enjoyable the sounds of the moveable element of water. The murmuring of a creek, the playing of the waves of a lake, they are always attractive for our ears and eyes, but **the fish pond enclosed in the interior of the forest, resting there solemnly,** has no less attraction. This has to be considered in the planting of the edge. There is no tree species which would not present itself particularly beautiful along the water, that means more beautiful than elsewhere; but every tree species has certain characteristics which show to advantage the most beautifully only near the water; so the alder is effective by contrast of form and color of the rising dark trunk, while the shining lights of the leaves fit harmoniously with the shining waves; spruce and fir as well make a contrast by the rigid growth of the trunk, while the story-like structure of the branches fit harmoniously to the horizontal lines of the water. If one wants to cause an effect mainly by contrast (it will be softened by the reflection in the water), one should choose the oak; if one wants to create really "idyllic" scenes, then birches, poplars and willows offer themselves, being related to the liquid element by softness and movability of their forms. So there is no lack of choices, and it should be used for real variety; but this is not to be interpreted in the way that every 50 steps another tree species should enter, quite the contrary. It has to be considered: two small ponds, **one of them** bordered mainly with soft deciduous species, **the other one** enclosed by spruce and fir, are quite different things, although they would give one and the same impression if they were bordered with a mixture **of the same kind.** Similar observations are valid for **coves of larger water surfaces,** and it has to be remarked, it is applicable for meadow edges and even for field edges as well.

The ponds themselves have to be livened up by **fish. Mosquitoes** and **horseflies** cause considerable discomfort for every forest visitor, but especially for the hunter sitting in the raised blind. Adequate regulation of the hydrological conditions may reduce this evil considerably. If swampy areas are partly drained for pasture cultivation, partly dammed up for fish-farming, the breeding grounds for this annoying vermin are narrowed down, and the fish are eager consumers of mosquito larvae. Especially successful in this hunt are golden orfen, and they show much to the viewer while they are hunting.

It seems to me, in my forest ponds, which appear dark due to organic acids, the golden yellow of the orfen is particularly beautiful.

Fields in the forest cannot be formed easily into a source of aesthetic satisfaction, not even with the best edge planting. They **often appear the most indecent by their very poor crops,** because the manure pile is rarely close enough to them. Types of management which supplement the insufficient animal manure by mineral fertilizer may sometimes be able to take remedial actions in this case, but mostly it will be best to exclude the undesirable area by a **clearly visible boundary** distinguished from the forest district as a body not belonging to the district. This may be done, depending on the circumstances, with more or less elegance by **game fences, hedges, embankments and ditches.** Especially for the land appropriated for use of officials, well-maintained fences and hedges are an adequate luxury.

Prince Pueckler characterizes the aesthetic value of **fences** very well when he says: "I have often found that a fence here and there, especially where the character of the region changes, has a very picturesque effect; yes, I even would like to say that it prepares the mind for new impressions and offers a calming section."

Fig. 41

In regard to the **living hedges,** the recent art of gardening design has often thrown out the baby with the bathwater. While once hedges were established just for themselves and formed into elaborate shapes, the hedges now have become too sober and straight. A hedge that has been planted for use is to be formed into a beautiful shape at the same time, for example by bordering the entrances with columns or arcades made out of living branches.

Much more beautiful than a clipped hedge can ever become is the edging of the forester's private plot in Klein-Ujeschuetz (Chief Forest District Katholisch-Hammer). There, mighty cypress-shaped junipers stand along the Trachenberger Line; but this picturesque hedge takes much space.

SECTION B. Embellishment of the Forest by Special Actions Pursued Particularly in the Interest of Beauty.

Chapter One

Park or Forest

I have been reproached as often as I have been praised for wanting to make the forest a park. I have always been more inclined to see a reproach in remarks like this, for I regard the park nearly as an evil, however sometimes a necessary evil. In the park I am disturbed by the idea that large areas are taken out of utility. As much as I stress that sometimes one wants to rest at a place where not everything smells after profit and percentages, on the other hand one does not like to have large areas of which one has to say: Here little or nothing is happening of the kind that the human would also be able to subsist on. Quite the contrary, the park needs careful and expensive effort and maintenance. If this is lacking, the park offers cause for justified criticism. Also, if one does not consider the costs, it will nevertheless **appear only rarely appropriate to lay out a park near larger forests.** Where the forest is not of that low quality, as Prince Pueckler found it near Muskau, there a **well-groomed garden and the forest, set beside each other without the mediating connecting link of the park, will show each kind of area to the most advantage.** If the forest cannot be extended directly up to the garden, then a connection sheltered against wind and sun can often still be established by a parklike landscape.

In the forestry literature it often is suggested **to treat parts of the forest like a park,** in order to satisfy the public in this way or to fulfill another need.

Recently three such declarations have attracted my attention, which I am going to quote here:

Weise puts **"park management"** between the regular and the irregular selection forest. He writes about that:

"Park management has up to now not been regarded as a type of forest management. With the increasingly critical need to have continuously shaded forest near cities and near popular summer resorts, the forester should acquaint himself well with such management and the aesthetic management of the forest, especially because the areal extent of such management needs to be only a very modest one and mostly a narrow screen will suffice in order to be able to practice any form of forest management behind it, untouched by the protest of the public."

"With park management the individual tree is supposed to cause an effect by the beauty of its structure, and the group is supposed to be effective either by its mightiness or by contrasts in shading and type of illumination. This can be achieved with the individual tree only by allowing it to grow up totally free from youth. The trunks united into groups need to be planted in such wide stands that each individual can achieve a full crown and therefore can show to advantage in the group. Besides deciduous trees, evergreen conifers need to be planted. Their landscape value, which is less outstanding in summer, soon comes more into the foreground with the colorful scenes of fall in order to be fully recognized in winter. Finally, the May green of the deciduous trees appears very effective in front of the dark background of the conifers."

I felt obliged to quote these instructions in their entirety due to the influential personality who expressed them, although I think they are right only in a small part, as the following pages are going to show.

Similar to Weise, Kraft also has expressed himself:

"The finished park also is similar to the regular selection forest, which is characterized by a spatial separation of the age classes in smaller or larger groups; it differs only in that it also contains beds, lawns, etc., besides trees in groups, which can be compared to the youngest class of the selection forest in a certain way, although they include as a rule a considerably larger percentage of the whole area than is the case with the youngest age class of the selection forest."

At the same time the **Prussian Upper Chamber** discussed the question of what a park-like managed forest is.

Count of Tschirschky-Renard had put in a request to declare the Forest District **Grunewald [forest in the city of Berlin] as a Federal Park.** The goal was "the formation of a primeval forest, established by nature and beautified by the art of the axe."

In rejecting this request, it was **passed** "to ask the Royal State Government to see that the Forest District Grunewald will be managed **in a park-like way in the interest of the public** and with special consideration to the saving of mature trees, and will not be decreased by sales."

The Minister of Finance von Miquel soon criticized the expression "park-like," because it is ambiguous and unclear. I will hardly be wrong if I suppose that the upper house may have thought of a type of management, about which Kraft writes:

"There is the actual park management and a second type of treatment of the forest property related to it, which has to be distinguished from it. The latter is supposed to save the actual forest character fundamentally, and is not directed to the **highest forestry technical exploitation** of the forest, but first of all towards the realization of forest aesthetic claims."

In the sentence mentioned above the emphasized words are the ones that cause offense for me. Weise as well as Kraft think of a type of management that causes trouble for forest management. This can and has to be avoided after all. As long as the demands of the aesthetician remain within forestry limits and the landscape gardener does not speak the last word, the "highest forestry technical exploitation" will not decrease. On the contrary, I see "a bloom of the well-ordered modern management in care of the beauty in forest," like a very benevolent *review of the first edition* of this book testified to me.

By a park-like managed forest district I would understand a district which is managed for forestry particularly circumspectly and diligently (in an elegant way), and in which the *landscape gardener* has the permission additionally to perform those actions he regards as desirable in an amount that the forestry purposes are not at all or only very insignificantly reduced by them.

The chances for friction between the forester and the landscape gardener will be excluded where both qualities are united in **one** person. This occurs not so rarely, and I found numerous creations of foresters with which the landscape gardener would hardly have found any fault.

One need not fear that the public looking for recreation in the forest could come off badly in that.

It had already been remarked very suitably in the commission of the Upper Chamber when discussing Tschirschky's request, "that it is much more interesting and educational for the population of Berlin if it can educate itself, in the sight of a forest managed in an economic forestry way, how to grow and maintain a forest, than to look at a miserable primeval forest."

One should not believe that laymen would not have a sense for good management even if they occurred in great groups of urban public. If a management unit is totally at its best, when it is, so to say, **elegant,** then this is going to be obvious to everybody. As the non-soldier appreciates a perfect marching parade, also the non-technician admires the useful movement of the steam engine; **it is noticed as well by the layperson if a forest management rises high above the average.**

The opposite of satisfaction has to come up, if one does not know whether one is standing in a forest or a park. The forestry rule will be condemned because it is not park-like, and a sacrifice brought for landscape gardening interests will be disapproved by the forester. The aesthetic demands of unity, of correspondence of appearance, and of and nature of the things cannot be fulfilled by an intermediate thing between forest and park.

Already Prince Pueckler has prescribed distinct separation. He wishes to distinguish the landscape garden from the park, and the park from the surrounding landscape by a clear boundary.

I saw a good example of such separation in the Grand Duchy of Oldenburg State Forests near Eutin in Schleswig-Holstein. There landscape strips, which were to be managed not for forestry but **only** according to beauty aspects along the parkways, are clearly separated from the forest by cut stones.

Separations like that correspond with the **far-reaching differences existing between forest and park.**

The forest is the more perfect the more it yields, but the landscape gardener is not allowed to strive for material gain. Some garden artists even have the principle of removing anything profitable, for example fruit trees.

How differently the forester and the landscape gardener have to work may be demonstrated by some examples: A selection forest, if it is **part of a park,** is to show as many as possible picturesque and different shapes of trees. The **forestry** selection forest is mainly to produce **timber,** unless it is only a barrier forest in high mountain ranges. Accordingly, there will be rather different ways of **handling the axe.** The forester will favor the straight, branchless trunks, he will cut out the gnarled, competing trees, and he will prefer tree species highly usable for timber before those less valuable for timber use. A forester will sharply attack seedlings of bushes, for example elder, while thinning in the selection forest in order to favor the establishment or sprouting of timber trees and to create regeneration groups as dense as possible. The landscape gardener is not in a hurry to do that. His beeches, his pines are allowed to stand for 100 years more than the ones in the forest—so why regeneration so soon! Under certain circumstances he may favor shrubs as an understory so much that he might chop young beeches, if they happen to appear by their own will, in order to save the European Cranberry bush, the holly or the juniper. The forester will aim at cutting out early enough the suppressing soft hardwoods. The landscape gardener will prefer every now and then the early blooming willows before the oaks, where he finds enough of the so-called "high quality tree species"—he will even be able to refer to Kraft himself in doing so, who recommends choosing "tree species with differing natural lifespans in order to establish the park groups."

The sharp difference presents itself in the layout of the roads. We avoid "lost incline"; the landscape gardener likes to bring more change into his route in leading it upslope-downslope, he loves to combine nearly level sections with steeper ones, which offer an opportunity to him to construct some steps. There are many more differences like that which could be listed.

Section II B of this book is therefore not an instruction for landscape gardening, but it is meant only to give some hints **on how the forest may be enhanced by some measures borrowed from the art of gardening.**

The relationship between the art of forestry and the art of gardening is similar to the one between architecture and sculpture, when the architect calls for the sculptor in order to adorn a building by his relief-friezes, by statues and carrying consols [corbels].

In this respect, the most perfection is achieved if, like Michelangelo and Schlueter, the artist is architect and sculptor at the same time. Schlueter's masks of dying warriors in the Zeughaus in Berlin are a particularly apt example for well-chosen visual artistic decoration. Michelangelo was even a painter, too, but nobody should suppose automatically that every forester knows at least two other skills, that he could be quite naturally also a garden artist.

This warning is not superfluous at all, for such a combination meeting of a double talent usually is presupposed as something self-understood. It can be seen with foresters who want to dictate rules to the landscape gardener without ever having even skipped through a single book on the art of gardening, and not less with the public, even with high administration organizations, which entrust without choice the first educated forester that comes along with landscape gardening tasks which might be solved only by individuals specially educated and trained for them. Usually then disappointment is not rare.

I do not write this in order to keep professional colleagues away from the care for beauty, for the whole meaning of this book is to inspire to beauty care. The discussions above are only to fight those voices which recommend the forester to reshape a part of his sphere of activity into a park, and which misdirect him to the assumption that he will be able to do so without the help of a landscape gardener. Sometimes the **relation of the forester to the landscape gardener** will be similar to the one between the client and the architect. This will, for example, be anywhere where public parks are under the supervision of the Royal Government and therefore actually are subordinate to the head of the forest administration.

In order to increase the reputation of the forestry profession, it may help a lot if the foresters prove to be suitable for this task.

Until now this has only rarely been the case. In this sense Wilbrand writes, complaining about mistakes made: "In the eyes of really educated people the forester will not be looked at seriously as long as he will not only avoid such mistakes, but until he will have assumed the leading role in regard to care of the beauty in the landscape, a role for which he actually has a vocation."

Chapter Two

The Enhanced Forest.

I call an **enhanced** forest a forest in which quite special attention and considerable energy and time are devoted to considerations of beauty, without significant interference with the effort directed to the highest net yield. The expression "luxury forest" which I used in the first edition does not satisfy me, neither do the terms "*Voluptuar* [Pleasure] *Forest* or Decoration Forest" used in Austria; for one can enhance without doing luxury, and the Austrian Latin-rooted words do not fit the nature of the thing either; therefore, I prefer to describe the term only with the words **"enhanced forest."**

Changing only the name I maintain what I wrote on the matter in the first edition:

Near cities or at spas, also in the nearest, well visited vicinity of country homes, it will definitely be advisable that the owner—not only the private person, but state and community even more—takes care that everything in the forest might be as beautiful as possible, and the beauty also might be accessible, and this in a higher amount than could be done over all the forest area, but **the management must not suffer from these efforts.** The economic management has to be allowed to go its own way totally without hindrance and according to its own principles, while the **owner himself pays for quite a lot of modest applied luxury out of his private purse,** for example, rather neatly cut compartment stones, pretty signposts, a particularly friendly forest house, an improvement of the roads and bridges beyond the measure necessary for wood transportation and beyond the needs of the forester, who is equipped with waterproof boots.

It is most important that one choose a **manager** for a preferable district like this who possesses understanding and interest in the care of the beautiful. His sphere of activity should not be measured too large, in order to give him time and freshness to keep himself free of the craftsman-like inefficiency by observing and thinking. The manager has to be paid adequately so that he will be able to work without worries. Money for educational travel and field trips has to be supplied for him.

But one has to take care not to get into a likely danger: Prince Pueckler put great emphasis on the clear separation of the "pleasure ground" from the park by a visible boundary, so that the park will never, not even for one moment, be able to appear as the badly-groomed continuation of the pleasure ground. We have to be even more **careful not to make any arrangements by which our forest might be suspected to be a badly kept park.**

If one stays within these limitations, the costs will never be able to increase considerably.

In order to give a certain figure I note that the "enhancement of the forest" costs 1,100 marks per year in the Eisenach Forest, namely 700 marks in the Forest District Eisenach, 400 marks together in Ruhla and Wilhemsthal.

However, in the amounts above, the main thing, the maintenance of the Castle Wartburg, is not included, which gives the value only for the area.

Where tourists are at work the forest owner will make the "Associations for Embellishment" [organizations founded in order to increase tourism and attract visitors] pay for part of the costs without thinking twice.

Like Goethe puts it as nearly self-understanding at some places in the next chapter, the enhanced forest should not lack a **building** that offers itself as a **goal for excursions.** Its size will need to correspond at least somehow with the extent of the area. There are numerous levels between the simple shelter and the hunting castle, and there is no lack of good examples. Adequate to the situation in Postel, my father constructed the "Johanna's Height" in 1849 and 1850 [see Fig. 42].

Fig. 42

In the discussions dedicated to the theory of the highest net revenue, it has already been proved that one has to add the yield of beauty to the value increment. This certainly is valid, especially in districts that are regarded as forest to be enhanced, and this not only in regard to measuring the rotation period, but also in choosing the tree species and the type of management. Leuer has presented an appropriate example:

"It is not unlikely at all that with reasonable prices for oak tanbark, the calculation has the result that the most advantageous management of the forests surrounding the city and spa of Baden-Baden is the oak coppice forest management. Now imagine if the forest administration would put that into practice. The delightful fir forest, which winds itself as a velvet green garland around the Baden valley and gives it its main attraction, would be cut down and on the bare slopes oak coppices would be grown, which are cut over every 15 years and then offer the barren sight of the clear-cut areas. Now, would this be possible at all, without provoking a cry of indignation throughout the whole of Germany and beyond its boundaries, without an unanimous destructive judgement of the educated world and without driving away the majority of the tourists who look for health and recreation in Baden-Baden and to whom the city owes its increasing wealth?"

The Grand Ducal Forest Administration of Hessen recently has decreed, taking into account opinions like these, that near the rising world spa Nauheim the very profitable oak tanbark coppice has to be stopped and that the oaks first have to be grown into a high forest, guided by the certainly true assumption that the indirect advantages will turn out as a rich replacement for the loss of the bark.

Otherwise one might feel obligated to accept the discomfort of the selection forest management for the interest of beauty. But I believe it would be wrong to recommend that in general. The demand of the public, possibly justified, to have certain paths always protected against sun and wind, can also be fulfilled in high-forest management when one regenerates naturally or below shelter and leaves abundant standards along the roads. Allees, to which a special chapter will be dedicated, serve the same purpose.

Fig. 43

Everything that has been said in Section A of this Part will need to be considered in the enhanced forest, but especially the advice for marking stands for cutting and for thinning. I know of two examples where the removal urgently necessary for the maintenance of the forest could not be done because the public, startled by early cut blazes, had written to very high personages and lodged a complaint. Removal in slowly advancing strips is particularly suitable to regenerate a stand without hurting the sensibilities of the public. The picture (Fig. 43) shows the "Susannen-Pines" in Postel which are converted into an oak stand by edge removal of strips after a Borggreve selection thinning had been done.

Regenerations performed in this way offer the advantage that they make the forest very opaque, which is especially important for compartments at the edge.

That much in general. The following chapters will go into more detail.

Chapter Three

The Establishment and Management of Park-like Landscape

In the first chapter of this section I tried to prove that the forester would be mistaken if he transformed a part of the forest area into a park. He is not to hand over anything to the landscape gardener; quite the contrary, he would be well-advised to practice peaceful conquering in moving his plantings forward into the terrain beyond the boundaries of the closed forest, thus establishing park-like landscape [open landscape].

If one demands a **definition of park-like landscape, I would like to say that it is usable landscape, decorated by woods, accessible by well-laid out but modest roads.**

Already von der Borch, the oldest German forest aesthetician, has rightly noticed the beauty value of the **smaller woods** distributed in the landscape, but landscape gardeners and poets have by far preceded him in its appreciation. I found probably the oldest glorification of copses when I leafed through a very old classical book in an hour of leisure. This was a German translation of Josephus, published in 1574. In the old style of writing here I quote his complaints about the devastation of the Jewish land:

"Although the Romans, with the planned construction /and also the transport of the timber /had great labor to do / nevertheless the truss of the bridge was erected by them within twenty-one days / also **all the forests and woods eleven miles around the city were cut down / therefore the Jewish country had become rather barren and unsightly** / which before had been adorned with green forests and nice gardens. But after all the trees had been cut down everywhere / it looked like a wilderness and was so devastated / that no foreigner / if he had seen before the magnificent country

XII.
Kratzkau. Blick in die freien Anlagen.

and the tremendous sites/ and now looked at the starvation / could refrain from crying and sighing **because of the change that had happened. For the war had taken away all the adornments and all the beauty** / and if someone / who before had known the country well / came there unexpectedly / he surely would not have known the place anymore / but would have needed to ask for the way like a stranger."

Modern art of gardening has cared right from the beginning to enhance entire landscapes by field woods, but it turned out that it was not easy to restore artificially what had been destroyed by indifference and short-sightedness.

When in England the natural style in garden design still was something new, people hurried to establish a little wood of larch trees right on the center of every hill; this then looked "like a little hat on the head of a giant."

In addition to the first masters of the English garden design and surpassing them, Goethe and Prince Pueckler showed better ways. Goethe did so in the novel ***Wahlverwandschaften*** [Chosen Relatives]. In this wonderful novel the fates of four persons, who are introduced early in the story, are as closely connected with the beautifying of a wide landscape as in Homer's *Iliad* the fates of the Greek were with the battle against Troy. One might want to recall the plot:

Eduard has taken care of the garden, but this is not a suitable job for the husband, for he should have left this to his wife ("the garden is the house extended into the outdoors," Prince Pueckler says, therefore in it the mood is allowed to work). At the same time Charlotte works on the park-like landscape. Now the captain arrives and brings home to both of them the point that Charlotte is not a match for this subject. Charlotte sees that he is right. At first a bit offended, she soon works together with him in a larger scale. Soon she falls in love with the captain. Now I skip the reshapings in the park, to which quite a few remarkable events are connected, in order to mention how it is Ottilie who instinctively discovers the only suitable place in the park for the garden house to be erected; because of that, Eduard falls in love with her even more than before.

Whoever cannot or will not really realize that park-like landscapes like that are truly a serious and important subject might want to read it again in the *Wahlverwandschaften.*

Also in the novel *Wilhelm Meister's Wanderjahre,* some valuable remarks are found upon this matter. So, in the novella put in there, called **"Who is the traitor?"** (Chapter Eight of the first book of the *Wanderjahre*), the very successful park-like landscapes of the "Oberamtmann" are described in detail:

"The Oberamtmann . . . had, after his own view and attitude, after the likings of his wife, at last after the wishes and ideas of his children, created and favored first larger and smaller separate areas, which, carefully connected little by little by plantings and ways, presented a most lovely, sometimes diverse, characteristic sequence of scenes in walking through . . . To the main buildings and working quarters, leisure gardens, fruit gardens and lawn gardens were added, out of which one came without really being aware into a small wood with a wide, ridable road, curving here and there, up and down. Here in the center, on the most significant height, a hall had been erected with adjoining rooms."

Following Goethe's ideas I here anticipate the piece of advice that **village roads** usually are important connecting elements of park-like landscape. Sometimes the influence of the respected forest manager or of the estate owner can do much to improve them and to enhance their surroundings. My father acted in this spirit when he had a wild pear tree, standing in the village meadow, fenced as a decoration for the school garden, with the explicit condition that its crown would have to be prevented from getting misshapen by grafting. The picture [Plate XIII] shows this tree decorated by winter hoarfrost.

Returning to general contemplations, I recommend park-like landscapes for **all regions with generally poor soils, all rough sites**

and all the sites which contribute to problems in agricultural use, such as hilly terrain, shallow soils, clay-loam veins, etc.; finally, especially for those sites where hunting is greatly enjoyed. But it would be wrong to believe that all so-called good areas have to remain closed to park-like landscape, that they would have to remain the sugar-beet deserts, which they often appear to be. Particularly on good soil the timber growth is also very favorable, and accordingly, wealthy villages in the near vicinity provide a good market for all produce yields that are offered by the composite forest [coppice forest with standards] (from broom twigs to the mill- wheel), and what a stand of roedeer, what an abundance of pheasants can be grown on 10 ha of woods, if 1,000 ha of the best fields and pastures accompany it!

The smaller the forest areas, the more freely the manager will be allowed to move. In **small** situations, it is possible to treat individually not only every stand, but nearly every tree, and to adjust to any changing need of the timber buyers. In this circumstance lies **a special advantage of management in the small woods of park-like landscape and these form the most rewarding area of forestry art due to that freedom.**

Whoever can get the **advice of an experienced landscape gardener** for layout and management of park-like landscapes should not miss this opportunity. In this way, **Kratzkau has been improved.** In 1848 my father took over this estate situated at the Weistritz below Schweidnitz. There he found a very magnificent residential building. But, situated between the farmyard and the woods in the swampy moat, it did not offer a pleasing view at all. The graciousness of King Friedrich Wilhelm IV (he knew the place from a brief visit on the occasion of a military maneuver) twice sent the Royal Garden Director Lenne, one of the first landscape gardeners of those days, in order that he design a plan for the enhancement of the property. Lenne's performance was a drawing for only a very small garden, but also an extended park-like landscape, which were soon started. Accurately following the artistic plan,

XIII.
Postel. Wilder Birnbaum am Dorfanger.

beautifully shaped meadows were created by land clearing, and on these, groups of shrubs and single trees were left at appropriate places, sometimes even supplemented by planting if necessary (a group of spruces, a clump of ashes). Later, nicely located paths and roads were built which served economic purposes at the same time. Their success increased the enjoyment in creating. Quite a few valuable elements have been added to the Lenne plan recently (by Roennenkamp) and have been developed by my brother. Now, if somebody who does not know the history of this piece of landscape art work rides through the area, he will say: How beautiful, how magnificent, here it can be seen once again: "Free nature is and will always be the most beautiful anyway."

This chapter was preceded by a picture that shows a view into the park-like landscape of Kratzkau [Plate XII]. The two oaks in the middle ground had previously been more beautiful, with branches down to the ground, but they covered the background. Due to an idea of Field Marshal Count Moltke, the lowest branches were sawed off close to the trunk, and now this landscape scene proves how much can be performed by sensible pruning, how large effects sometimes may be caused by small means without costs, if the details are treated with consideration of the whole.

Whoever doesn't have the opportunity to consult a landscape architect, due to lack of finances or other reasons, will create something good, too, if he proceeds slowly and avoids schematism and has a detailed knowledge of the terrain. But whoever, abandoning the secure philosophy of usefulness, might want to plant **every** hill with trees, to veil **every** bit of water with willow bushes, to border **every** road with mighty rows of trees, who might want to straigthen **all** the curved roads or to curve **all** the straight ones, in doing so he would not act with less understanding than others have often done, but he would miss the stated goal.

Also someone who consults the landscape architect must have already carefully considered the task he wants to set out for him, therefore he needs to have some understanding of the subject. I think it is useful to put in here **some principles,** whose consideration can prevent quite a few mistakes in establishing park-like landscapes:

On the area available one should pick out all the places where agriculture is not practiced at all, or else not to a sufficient advantage; that means besides all the already existing wooded parts and old clay- and marl-pits, all the very poor cropland areas, swamp holes, and the shallow and steep field parts. One should visualize these covered with tree groups and ask oneself, What impression will this cause? After that, one should sort out again the ones whose planting only would create disturbance for the entire scene, and continue to torture oneself further in practicing agriculture on them; but the others should be taken into adequate forest use. If then it turns out to be desirable to give a better shape and an adequate connection to the tree groups, then one might want to sacrifice a little piece of better field land to this purpose every here and there.

Not only must one pay attention to a **layout** as favorable as possible, but one must also see to an advantageous **structure of the groups.** Therefore it is recommended to use plants of uneven height from the beginning, but the group should not always be highest in the center; instead, one should remember the **Golden Section** as it was taught in Part One. **Nature offers excellent examples** of how trees and shrubs have to be grouped in a beautiful way according to kind and height; one needs only to diligently train the eye in the art to test **impartially** such patterns and to appreciate them.

In difficult cases I have helped myself by putting several cart loads of two to five meter high birches and pines into the ground, which had been obtained while thinning and weeding, in order to view and be able to improve the effect of the planned plantings. Whoever is better trained than I am will achieve the same result with only some bundles of straw.

Sometimes landscape gardeners are in the comfortable situation not only to be able to use the terrain, but also to reshape or even to create it, as was done, for example, for Prince Pueckler in Branitz and for the Duke of Braunschweig in Sibyllenort. For park-like landscapes, similar sacrifices need not be made; here it is only our task to **make the characteristic qualities of a landscape visible in the most beautiful way,** the heights as well as the vales. Some simple examples will show better than a long essay how this is to be understood:

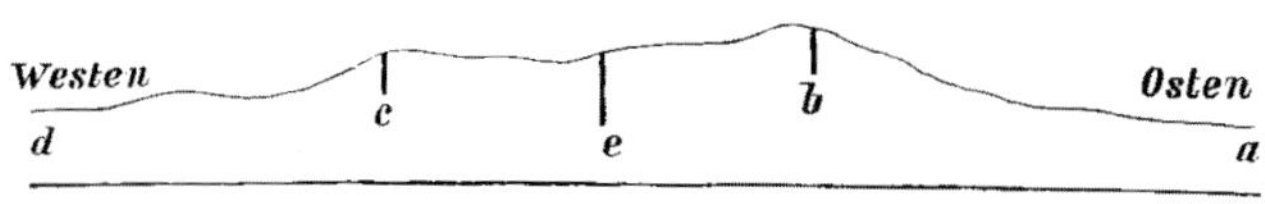

Fig. 44

Suppose there is a sand dune in a region usually rather level, appearing in the long section like that shown in Fig. 44, and it is planted with even-aged pines. There it is recommended to lead the first cutting up to b, then some years later the second cutting up to c, later the last one up to d. At these points the steep tree wall at the beginning of the cutting and the stronger angle of inclination of the hill support each other's effect. Quite the contrary would be true if the cutting were done in one year up to the center and after that until the end.

Furthermore, an elevation difference of a few meters often changes the total character of a property in the plain. The low sites are preferably left to cultivation of pastures and meadows. Under those conditions, forest edges along the lines where the terrain changes will always have particularly good effect. Of course, they do not need to form an uninterrupted continuous border; quite the opposite, there must be views from above down to the meadow.

The **lowest** areas are marked by the routes of the drainage ditches. Even small ditches, carrying water only temporarily, can gain a certain importance for the landscape if their borders are grown in a beautiful way. The most obvious characteristic of the ditches is their winding route which stands in the closest cause-and-effect relationship with the tree and shrub vegetation, for often one can see the creek taking a totally changed direction, after being obstructed by a single old alder root clump. There it is important to adjust wood growth and watercourse to each other by maintaining their essential relationship, by removal of more coincidental things and by sensible completion. Particularly **in the plain** where these examples have been taken from, often very small means are sufficient to cause great effects. The opening-up of a view, the release of a large tree, the establishment of a group of trees where the existing buildings and tree groups lack a correlation, often can contribute more to the enhancement of a country house than could be done by gardens costing thousands of Thalers. This is by far not sufficiently recognized; often even the contrary is done by ditch course changes.

Particularly in the richer regions one can often perceive that here and there a wealthy man on his estate has enclosed a rectangular piece of land with a fence or wall and established this as a **garden.** Densely planted all around, this garden does not offer even the least insight into its interior; it actually seems to be closed in a hostile way towards the outer world. Often the **churchyards** lie even more isolated in the parish land; I have been told that in Schleswig-Holstein in good areas even the woods used for timber are closed off in the same way. Embankment and moat enclose them, and barriers block the access. In such an isolation, then the tree world does not appear as an accompanying adornment of the landscape, but as a **strange,** gloomy addition. This unfavorable impression can be softened or removed if a connection of such isolated plantings is created.

In Schleswig-Holstein, this is done very often by hedgerows, but it can also be done by allees and other plantings. A special chapter will be dedicated to the allees. In regard to the hedgerows, I repeat my criticism that they screen the area too much and make it unclear. At least in some sections they should be replaced by low-trimmed hedges, like the Grand Duke of Oldenburg has often done on his Holstein properties in order to keep beautiful views open. **Where the land is not so valuable** that one has to be very economical with it, or where the **hunting use** promises a substitute for part of the estate's lost revenue, there some other development is recommended instead of the allees, as was planned, and even partly realized, by Prince Pueckler in Branitz after English examples; therefore, I will call it the **Pueckler**-hedge. He writes about it:

"On both sides along the road a sometimes narrower, sometimes wider strip, according to the situation, is trench-plowed and completely planted with young plants, like a forest planting, but in between individuals, higher groups are distributed, which form a sort of irregular allee above the understory. Where the bordering land is not my property, I content myself with continuing these higher groups by themselves, without additional planting, narrow along the road's edge."

"The young wood usually is treated like understory shrubs and it is cut every 6 to 10 years, but the higher trees are allowed to continue their growth."

"One can easily see that in this way even a poor region will soon gain a more friendly appearance from the road, while later on, by various treatments one can still create a lot of different effects, by letting larger masses grow up, pruning of single older trees, clearing out other individual trees, etc. Finally, one can always completely screen the disturbing outer landscape, wherever this is unattractive, by a welcome dense wall of foliage."

In some areas, where people do agriculture with a sort of fanaticism and think of it as worth every sacrifice, however, for a long time high, roughly-broken granite stones have replaced the street trees. Pueckler's suggestions are not meant to be for conditions like this. There one has to be glad if **at least every here and there a single tree** has escaped destruction. Writing this, I think of an old willow in Rosenthal near Breslau. It is not at all a particular large and beautiful tree, but it is the **only** tree there, and the eye, when it searches for an orientation other than that which individually sprouted beet-stems can offer, roams up to this willow. What an abundance of memories are connected to this tree for master and mistress, farmhands and day laborers! After all it is the one which serves as a sort of symbol for the adjacent valuable fields; it plays an important part in the bird hunt (the breakfast is to be brought there); the field workers rest in its shade; it serves for orientation for everybody.

But I have walked away too far from the forest. In some readers' mind already the reproaching questions may appear: What is all this? What do park-like landscapes have to do with forestry at all? I have to admit indeed that this is a controversial field. The landscape gardeners already want to annex it, and the agriculturist (it is open fields that are to be enhanced) is also going to join the discussion. The agriculturist and the garden artist—one always thinking of net profit, the other one almost never—will probably not easily be able to come to an agreement unless people with forestry-educated attitudes assume the role of intermediary.

One will be able to regard the park-like landscape as successful if it is often true what Goethe says about the creation of the Oberamtmann and his daughters: "It was indescribably beautiful! One thought immediately to have it seen everywhere, but nowhere so significant in its simplicity and so welcome."

Such aesthetic gain is often without cost, but mostly it can be achieved with very little cost. I estimate that 10 Thalers, spent for park-like landscapes contribute as much to the embellishment of a property as 100 Thalers for a park and 1,000 Thalers for a garden.

Chapter Four

Forest Enhancement by Establishment and Adornment of Roads (Crossroads, Roadsigns)

While in the first section of this Part, the decisive points for the layout of the forest road system have been developed, it is still left for discussion which further actions concerning roads can appear desirable from the purely aesthetic standpoint.

This can be two different things. On one side it can be the **increase in the number** of roads, and on the other, their **enhancement.** It may be appropriate to establish some other roads besides the forest roads and inspection paths in the interest of beauty, in order to make accessible those beauties of the district to which the existing roads do not lead, and in order to guide the public traffic away from the timber transport roads. It may be appropriate to establish special footpaths, horse-riding paths, cycling-paths and roads for vehicles, to keep all separated amicably.

Every regular park should contain a "by-pass road" which leads along the major beauties. On these roads the hiker should get the impression that all these marvels are presented mainly for him, for his pleasure and joy. It is recommended to wealthy owners also to construct a homogeneously shaped street in the forest for their own pleasure and for the use and well being of other people. Some of these exist near Wiesbaden. There they are called **round trip roads.** The road has to be able to guide the hiker without a road-sign.

Besides that, there should not be a lack of **smaller secondary roads and paths** which lead to separate places, especially in an area where many people and people in need of a rest unite; for the suffering human, as well as sick wildlife, often wishes to separate himself from other humans, to walk by himself and to sit down by himself. Therefore there also have to be seats. **Benches** under spruces and beeches offer the longest protection against rain. Under beeches the hiker may even wait for the end of a thunderstorm without fear of being struck by lightning.

Too many benches and roads will disturb the peace of the forest scene. There should not be established more of them than necessary, and they should not be made to stand out conspicuously by the type of maintenance. The type of maintenance has to make a visible difference between main and secondary roads. Of course, it never hurts if roots and stones are carefully removed and holes and furrows are leveled, but not **all** of the secondary roads and footpaths should have distinctly defined edges, not **all** of them are to be colored with bright gravel, not **all** of them are to be kept free from leaves and growing grass; otherwise the term of spa-forest will develop which does not fit into the forest at all.

Newly developed roads are always very conspicuous anyway, and where many roads have been constructed at the same time, they make the forest into an object for uncomplimentary comments for one or two years. Then it is said that the "many roads" have destroyed "all the forest poetry." This can be cured easily and quickly by **sowing grass on the fresh subgrades.** Expensive grass seed is not necessary, but the hayseed from the hayloft of the head forester's horse stable will perform excellent service for this purpose, for there is no problem if along with the grass some weeds such as *Plantago* become established. This procedure, tested and approved in Postel for years now, has some advantages also from the hunting standpoint, because of **browse for wildlife,** but of course I recommend it only for unpaved roads.

If the principles mentioned above are followed and the secondary roads are given only a moderate width (3 to 3 1/2 meters), then one can establish as many of them as necessary without worrying. Sheltered by a canopy of leaves, grown with grasses and other plants, or covered with forest litter, they will hardly be obtrusive. Such roads also can be partially hidden if they are broken several times so that the eye can follow the direction of the trail only for a short section. By the way, this is very useful in regard to hunting. Deer like very much to stand and browse on roads like this where they cannot be seen from a distance, and the host of the hunting event may excellently use the curves in posting for the safety of his guests in group hunts.

Caution in road layout has to increase, the smaller the terrain sections are whose beauty is to be presented. If a gently sloping mountainside is interrupted by a level road constructed in the middle of it, then the total impression of the terrain can be changed easily. Someone with narrow, deeply cut ravines in a plain should not at all quickly give in to the temptation to develop a path down there along the creek's bank. I have noticed several times that such small conditions do not tolerate even the construction of a small footpath without losing some of its aesthetic effects. Therefore it would be better to place the path up there at the edge of the top of the creekbank, and to make it cross the ravine by a wooden bridge at a particularly beautiful place, in order to offer a view up and down the creek for a change.

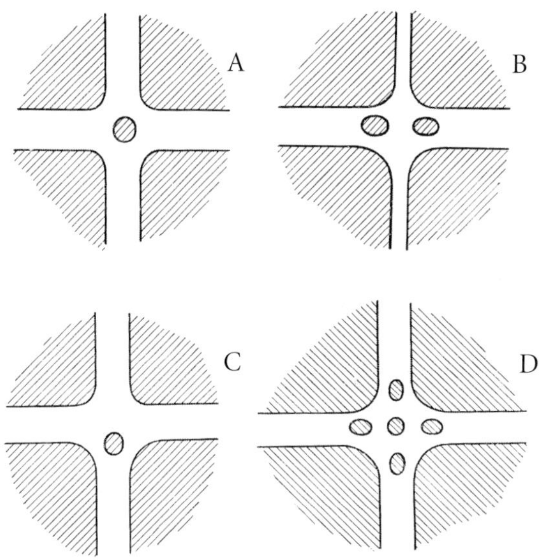

Fig. 45

The management does not enjoy the same freedom for establishing the main roads as it does for the secondary roads. Mostly the location of these will be predetermined, but one will be able to do small corrections, and at **crossroads** these can do a lot of good. Burckhardt has already recommended **"to fill the long and uninteresting tracks of pine forests with well-groomed groups of friendly tree species at the cross-roads, and to allow vehicular traffic by rounding the stand's corners."** I think he means like the Figures 45 A to D show, but this can also be done more easily.

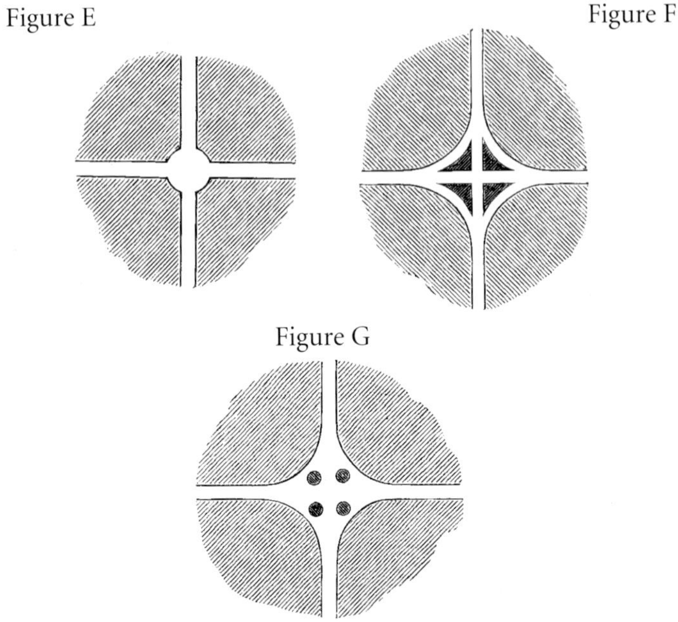

Fig. 46

Even the simple Fig. 46 E can look quite good in the forest district. Figs. 46 F and G show some shapes that are also acceptable, and many similar ones could be added to these. As with rectangular crossroads of only two straight tracks, the imagination has left open for it quite a bit of leeway. Even greater varieties offer curved ways, offset intersections, the branching of one road into two arms, the junction of several roads into one star of roads. Particularly varied terrain allows the most different settings by branching in the curves, and falling and rising road sections. These can increase their special attractions by appropriately uncut standards, and, if these are lacking, by a tasteful bordering planting. The number of possible, even the number of recommended designs of intersections is very great. **Even in larger forests, therefore, it is not a very difficult but a very rewarding task to give a special character to every single**

intersection; one only should avoid any "too much." Once I saw a small carrefour [rectangular intersection] that had been marked out according to the simple Fig. 46 E, but had been leveled very carefully and had been bordered with spruce. A forester's eagerness for embellishment had created it, while the approaching firebreaks were in a barely drivable state. Because of this, the creation, being commendable in itself, caused a rather unmotivated, nearly tasteless impression at that time.

The "filling" must not be done in the way that the view is completely blocked. The "friendly" tree species has to make the scene attractive, but it must not block the view.

In the Berlin Zoo one can find two excellent examples that show how to do it and what to avoid. For the first example, I recommend for consideration the dark yew at the Flora Square, which, from the distance, appearing in the center of the allees, offer a very attractive view, like Plate XIV shows. In the forest one would plant a pyramidal-growing juniper instead of the clipped yew.

In front of a dark background, a bright tree looks better. The small drawing inserted shows a Sessile Oak decorating an intersection in the pine forest (Fig. 47). The surrounding planting at the bottom of the Lion Group near the Brandenburg Gate is incorrect. Grown up too much, it "fills" the "steep area of maple" in an undesirable way.

Old trees will present themselves as a natural adornment at crossroads; **they will never really spoil a scene.** Once I spent a whole day of work in shaping two rather long firebreaks in a way that three particularly picturesque pine standards would end up standing at the corners, and I can say that this effort rewards me daily, whenever I come into the forest. Such trees unfortunately cannot be created all of a sudden there, if they are lacking; it is easier, then, to set up some **large stones** as randomly spaced as possible. These may serve additionally for putting an inscription on them, or even as a **trail sign.**

Fig. 47

Particularly for setting up trail signs, the center pieces in Burckhardt's pattern (Fig. 45 A-D) will offer an excellent site, but they also have to show themselves as somehow deserving in their furnishing for that preferred site, because after all, cut stone columns (the highest quality luxury) will always have to remain a rarity. **The inscription** ("Oh, how much do I hate black and gray! Less, white and yellow and blue," sings von Wildungen) should be yellowish white on dark green or stone-colored background and it should be written in a nice readable way.

Being the source of often unpleasant surprises and unexpected frustration because they often create more doubts than they solve, **the trail signs are quite able to include some sense of humor.** So one of them is very famous, because, in the middle of young pine plantations abundant with hares, it stretches out the silhouettes of five fleeing hares instead of arms; another one maintained its relation towards the public with this inscription:

> *"I am directed to guide the way,*
> *but not obliged to go with you."*

The humor in this is that the sign has been stolen nevertheless. Jokes like this, which anyway are allowed to occur only very rarely, will probably be the most possible where the hiker, being bored by monotonous conditions, will gracefully accept any kind of stimulation.

One should **by no means,** as it is often done, nail written signs **onto living trees** out of misdirected thrift. It is always embarrassing to see a living trunk hurt by a rough nail.

Neumeister says in this respect: "It certainly is a strange phenomenon that the dedicated guardians and care-takers of the forest, the foresters and hunters, do not keep themselves free of damaging the wood stands, but often even systematically support it. If one walks through a forest, one often sees the trail signs and warning-boards and number-signs of the compartments nailed or screwed onto the trees. It needs no proof that in this way many and often particularly valuable trees are damaged conspicuously, and mostly at a place which lies in the most usable part of a trunk, according to experience. . . . For curing these failures, there are some means given below, which have already been used to advantage in some forests for years:

1. Trail signs, prohibition signs and orientation signs, like for example for the numbering of the compartments, must not be nailed or screwed onto living trees. Rather they have to be fixed onto peeled stakes which have been charred and tarred at the bottom and have been driven into the ground at these places.

2. In order to save stakes the compartment number is to be written onto suitably standing trees with white color at an adequate height. In order to achieve uniform figures, it is recommend to

XIV.
Berliner Thiergarten. Allee am Floraplatz.

make stencils out of cardboard which are put on the actual spot of the tree and painted over with a paint brush carrying white paint. The paint is to be mixed out of white lead, varnish or oil and turpentine. The spot on the tree to be numbered has to be smoothed beforehand with a stiff scrub brush or by light flaking of the bark, which will at the same time provide a better background color."

"The entire costs of a unit numbering like that, including the loss of time due to the necessary travel from place to place, total at the most 10 pfennigs. Experience has taught that numbers like these stand for many years and do not need more expenditures for repair than the number signs."

Where the local Associations for Embellishment are at work, the colored marking of the roads has often exceeded any measure. Complaints have been made, and rightly so, that "on every tenth tree a blaze of several colors of the brightest shades has been painted there, with which the scene is spoiled and the eyes are tortured, so that walking between these screaming paint marks, some right-hand, some left-hand, is equal to running the gauntlet, regarded under the aesthetic point of view."

The eagerness for guiding must not go that far. As pretty as it is useful, I think of the marking that marks the way to the Kalbe on the plateau of the Meissner. I have been told that there, six-sided columns of basalt have been put up.

The following chapter teaches how to mark roads and trails by planting trees.

Chapter Five

Planting of Trees along Roads and Firebreaks

The previous chapter has been dedicated to the enhancement of the roads, trails, and ways, and this one is intended to serve the same purpose, but its subject is so important that it needs an independent treatment, for **the enhancement of the roads and firebreaks by planting their edges** belongs to the most efficient principles of care for beauty in the forest. By these there can be done much good, but a lot of harm also can be done. Harm in the respect **that it is by no means appropriate to emphasize every road by a special planting.** I repeat it (for such valued readers who only skip through single chapters, it is also said at other places): The **more a road is distinguishable as such, the less one will feel that he is in the forest;** then one is just on the road or even in the street, and not at all within the stands. Just in this feeling of being densely enclosed by the forest, there is a special attraction that, for example, is also experienced by the non-hunter and non-bird-trapper when he walks the hunting path or checks the bird snare. **For this reason I would like to see allees in the forest only there where a road by its width hinders the impression of forest enclosure anyway, or where more important sequences of roads, like the roads leading to the chief forester or to the hunting castle, need to be particularly distinguished, or along straight firebreaks and cleared lines as long as these separate the administrative units from each other and therefore also the different age classes.** But never should the "roads of the 4th order," leading into the interior of compartments and districts, and intended **only** for access to the individual units, be bordered and distinguished by a regular planting.

Also with roads which actually can bear the planting of an allee, one will need to be careful at places where they cross a pasture, so as not to interrupt the view across the meadow in an inappropriate way. This can be avoided by irregular placing of the trees, or by the use of high- and low-growing species, changing here and there also by the foreground-planting of groups, **more safely by complete transition to the most random road planting, to the "Pueckler-hedge"** which we learned about in the third chapter of this section. Often one will have to do completely without a road planting at places like that.

The warnings above seemed to be necessary in order to hinder any overeagerness; I hope they will suffice for that purpose! Now I would like to go over from the negative (how not to do it) to the positive.

I distinguish **two classes of road borders; first, those that stand to the surrounding landscape or to the adjacent stand as close as to the road itself, and second the actual allees with tree rows. The first class** is the one that includes the often-mentioned **Pueckler-hedges** and the **linings of stands.** Although allees can be grown out of these linings, they make quite a different impression in the beginning; therefore they are slightly less subject to the doubts against allees, mentioned in the beginning. **For linings of stands are plantings which distinguish themselves from the forest only by the chosen tree species, but are part of them by their position,** like Figure 48 explains better than many words. It is self-understood that the tree species preferred by edge position have

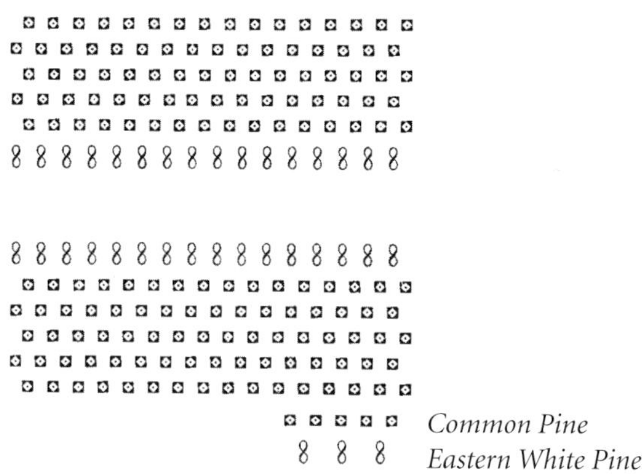

Fig. 48

to appear, so-to-say, more grand than the ones dominating in the stand, therefore one would not be allowed to line, for example, deciduous trees with spruces, but to line pines with spruces. Spruces themselves can receive an edge of firs.

The patterns of Figures 49 and 50 are particularly suitable for the embellishment of pine stands by deciduous trees. With positions like that, it is possible for the second rotation to save the oaks, or whatever the preferred tree species is, partly, and then in this way gain an allee so mighty right from the beginning that it is going to silence by its beauty all objections which might be justified otherwise.

The allees with tree rows in the more narrow sense, the planting of trees which are to appear as belonging more to the body of the road than to the neighboring forest, can be separated into two classes, according to whether the road trees have been planted with narrow distance and form a connected canopy of leaves, or whether they show themselves individually, being more distant from each other.

1/500 [scale]

Young Oak Trees in Seeded Pines

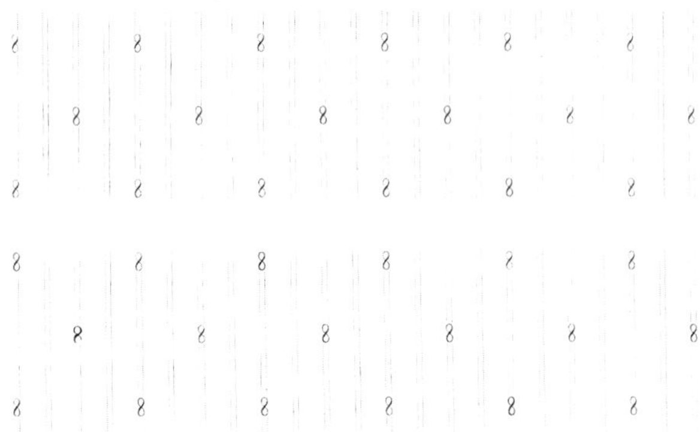

Fig. 49

The allees of the first type—I would like to call them closed allees—can gain the most marvelous effects with somewhat favorable tree growth. In order to secure this effect, one should limit oneself to only one tree species, and should choose, if possible, a species whose tops tend to come densely together, like linden, horse-chestnuts or beeches. The trunks should **not be set further apart than five meters in the rows.**

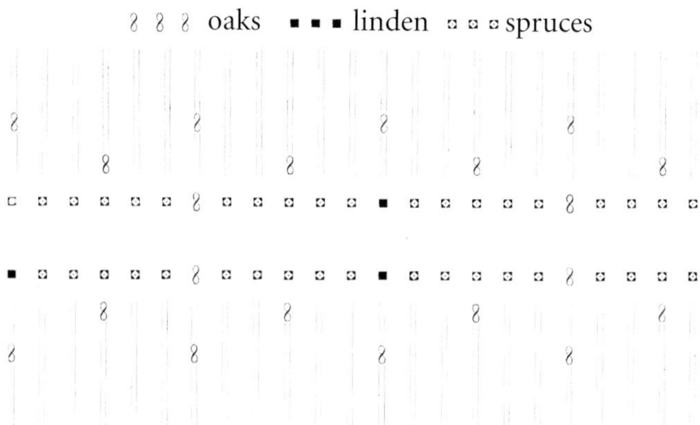

Fig. 50

The magnificence of the effect is increased considerably, and some comfort (especially a certain feeling of safety) is gained by the way, if on both sides of the main road, or at least on one side of it, footpaths are constructed and are planted in the same way. **Because every closed allee is embellished by regularity exercised as much as possible, it has to be seen that the trees are set along the string in every direction, not only along the long direction.** This can easily be done with the square-measuring used in plantation treatment. It should be marked out like the whole road area, including the footpaths, was to be transferred into a plantation with square planting with a planting distance of 120 cm. After that, nothing is easier than to choose the suitable planting places for the young deciduous trees so that everything will fit. Whoever knows how to handle graph paper, however, can plan it even more comfortably from the desk. It may be understood that even **ditches** running beside the roads sometimes may offer a reason for establishment of fourfold tree rows. Figs. 51 a. and b. may serve as examples in regard to straight arrangement also along the direction of the diagonal; Fig. 51 c shows how **not** to do it.

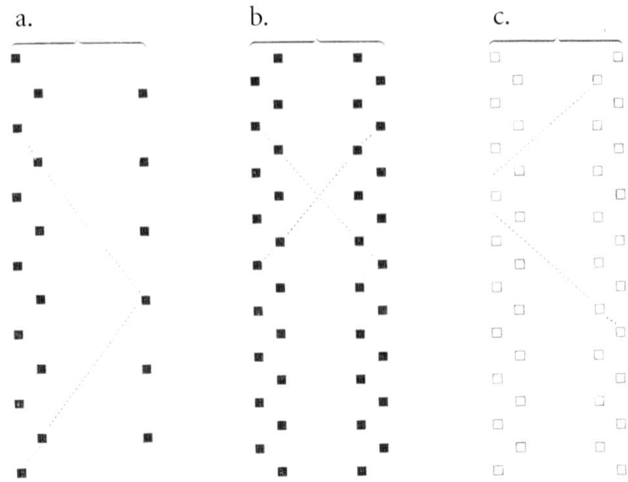

Fig. 51

It is bad that the road will dry out less, due to the severe shading, as the allee trees grow larger. For this reason, the **open allees** are preferred for many situations. These have to be planted and maintained (by early removal of trees) in a way that one tree crown is never restricting another, but that every tree can be seen and appreciated individually as an entity standing for itself. Also, open allees may be planted in one tree species for longer sections, particularly in places where much else can be seen, where the allee presents itself somehow only as a secondary element of the country road, for example in a nice mountain area. Allees of one tree species that extend from the valley floor to high up on the mountainside have the particular attraction that they serve as a scale for the mountain climate. I remember a road with mountain ashes whose fruits, already red in the valley, showed all the shades through orange and yellow up to green when climbing up the mountain. Also, where it appears to be appropriate to bring an open allee almost to the magnificent effect of a closed one, this goal can be achieved only by limiting to one or maybe two tree species that fit particularly well to each other, and this can be done most safely with **pyramidal trees.**

Unfortunately, the Lombardy Poplars [Pyramid Poplars] and the incomparably more beautiful Pyramid Oaks have the disadvantage that, if set up in long rows, they hinder the overall view of an area like a grid or wall. Therefore they are suitable only where not much can be spoiled with the area, or where there is no overall view anyway. An allee of pyramid trees must not be too short, otherwise there will be no mass-effect; on the other hand, also not too long, at the most three kilometers, otherwise the monotony of the homogenous trunks will get boring. It does not fit into the high forest, because the side shade makes the trees bald at the bottom and unsightly. For a **connection between a settled place and the forest,** pyramidal trees are more suitable, because houses and forest give good **ends** to its rows, which an allee can less afford to lack, the more imposing it is.

It is an often rewarding but always difficult task to combine an allee out of **several tree species.** By no means can this be done without concept. In selecting species, one should always think of how the tree rows will look seen from the side, in order to give a nice profile to the avenue itself; on the other hand, it should not screen the overall view across the landscape by tall tree crowns just at an undesirable place.

The further the trees are from each other, the more arbitrarily the selection may be done, but with narrower spacing (less than 10 m apart) it would be good to use a certain system. Here are some hints for that:

If the preferred tree species have different life spans, one must be sure that the species with the shorter probable life span will be juxtaposed with the longer lived species, so that after removal of the former, a regular pattern is left. I would like to recommend as an example Fig. 52 a.

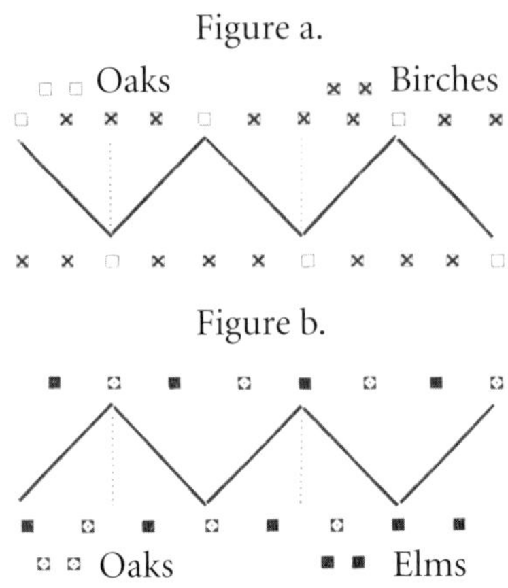

Fig. 52

A distribution **diagonal into the gaps,** as shown in Figure 52 b, has the disadvantage that an exact regularity will not result when some of the trees are removed; then one would have to decide to remove two neighboring trees at the same time and to keep the same proportions of the mixture as before, which, of course, is possible for a long period only with appropriate choice of the tree species. The trees remaining until last are connected by lines in the figures.

The planting will have to be the more varied and the richer, the more often and the more slowly the road will be walked along, that means on roads near the home and where sand or gradient of the terrain forces one to drive slowly at walking speed. **Species combinations that show something nice to the eye each season if possible** are particularly advantageous. Therefore the planting should unite the oak, which becomes green late, with the early birch or mountain ash; on the other hand the birch should always be mixed with tree species which are the most beautiful in midsummer (for example *Robinia* or mountainash). Deciduous trees are allowed to alternate even with conifers. The spruce fits particularly well to the linden, but not well to the oak. Always tree species have to be chosen which are at least aesthetically equal to the one in the neighboring stands—so the aspen, for example may be allowed to find a place as an allee tree in the pine forest, but not in a mixed deciduous composite forest. The number of permissible combinations is nearly infinitely great, and therefore it is impossible to discuss these in detail; I prefer to give some space to an example out of my own practice instead of a long explanation:

I found the driveway up to my house as a sandy road, some of it rising relatively steep, lacking a clear boundary, and trimmed irregularly with old lopped poplars and rotting moldy aspens.

This road, now appropriately regulated, has received an allee of linden to the edge of the village, in fact as an open allee of six different species of linden. Then, along with the forest, there are English Oaks in a straight line up to the first bend of the road, which row is ended by several Pyramid Oaks. Further down, where the road is bordered partly by someone else's fields, where therefore damage by shading had to be avoided as far as possible, there are *Gleditsia* standing alternately with *Sorbus,* the latter being of seven different species, but always two of the same species standing opposite each other. The *Gleditsia* are not a fortunate choice in that respect, because one must be constantly vigilant that no branch grows into the road, lest its thorns do damage. The popular rich variety always pleasantly entertains me when I drive home the distance up the slope at walking speed. In the forest, however, I have been able up to now to resist the temptation for a colorful mixture. There, I have never mixed more than two tree species with each other (oak with linden, oak with birch, oak with maple). Only for adornment of the intersections have I added at the most a third species.

There is another example from my neighborhood; this one is not totally out of reality, but arranged a little bit ideally.

For a length of about twenty kilometers, a country road connects two noble manor houses, each one leading directly to a town (Trachenberg and Sulau). Near the castles and the towns there are densely planted double-row allees of linden and horse-chestnuts for the evening walks of the townspeople; then they are followed by fruit: cherries for one section, then pears, finally apples, the latter all of one and the same species, all "Red Reinettes," which is known by the name "Purple-red Cousinot" among apple growers and recommended by them. This apple species is recommended for country roads due to its vertical growth of the crown. Its advantages are great hardiness against severe winters and late frosts, very high productivity, magnificent appearance of the red blossoms and the fruits, which, however, cannot be eaten fresh from the tree.

One should not claim that fruit trees are always unsightly. Whoever assumes that is probably thinking of shapes like those created by the disastrous "pot-pruning" of the crowns; but this practice has long since been rejected by fruit growers. Figs. 53 and 54 show how a well-grown fruit tree is supposed to look in its youth. Forms like these also guarantee a favorable development of the crown for later. The tree in its winter state shows how it is to be pruned, and the other shows the development in the following year.

A fruit allee combined like that will delight the eye at five different times of the year, by cherry blossoms, pear blossoms, apple blossoms, ripening cherries, and ripening apples.

Midway between the two towns there is forest, enclosing ponds and pastures, partly fenced and richly stocked with wildlife, with good stands of pine, spruce, oak, beech and alder. Maple has been chosen as an allee tree for the whole distance in the forest, and this in such a way that every other tree on each side is a maple, partly Sycamore Maple, partly Norway Maple. Always between two maples, there is a tree of another tree species, either because it fits particularly well for our site or because the choice is dictated by other circumstances, e.g. horse-chestnut as browse for deer in the preserve, ashes and elms at the ponds and in the pastures. At bridges, or where secondary roads turn off, linden stand between the maples. At these places the ordinary forms of the maple have been replaced by a magnificent variety of the Sycamore Maple which possesses a dark red color on the undersurface of its leaves.— So much for the example.

Fig. 53

Fig. 54

It is regrettable that injudicious treatment so often turns the aesthetic value of the allees into its opposite, namely that particularly the deciduous trees have to suffer so much by cruel prunings. **Nobody should establish more allees than he is able to take good**

care of. The growth of those branches, which tend into the road, is to be hindered by moderate cutting of their tips while the twigs are still young, so that the ugly amputations of larger branches some years later can be avoided. Detailed instruction for tree care by pruning has already been given above.

Allees suffer much damage by the actions of the telegraph companies whose demand that their wires should be safe against touching by tree branches is justified. Pruning of trees like I recommend will mostly lead the crowns of the trees quickly beyond the area of the wires.

Recently an application of insulation of the telegraph wires has been recommended; hopefully this method is going to succeed! It would be very important for fruit allees along roads.

For all allees, but especially for the "closed" ones, a regular development of the little stems is important. Therefore it is recommended to plant some young trees of the same kind at an appropriate place (**tree nursery!**) simultaneously with the establishment of the allee, in order that they can grow up with the allee trees and can keep up with them and will be ready as an **exactly fitting replacement** in the case of need.

Allees that are to appear soon as "closed" have to be planted densely, and therefore it cannot be avoided that the crowns interfere with each other. If removals become necessary because of that, it will take several years before the allee will be closed again and will make a good impression, as before. In Nauheim, Wilbrand showed me how this latter evil can be avoided: There, in an extended allee, the tops of every second tree had been moderately cut. These trees, now disadvantaged in the fight for existence, were overtopped by the others, which closed themselves together above them. The removal of the cut-back trunks may be done in a few years from now without leaving behind noticeable gaps.

The forestry rules discussed in this chapter and in the previous one had the common advantage that by them and with some good intention and moderately favorable conditions, the total look of a district can be improved within a few years. Whoever has worn a military uniform will remember how much a new edging of the collar and well-polished buttons helped not only the garment but gave a clean and tidy impression to the whole man. The adequate adornment of the forest roads makes up as much for the look of the forest district; but as well as buttons and collars do not yet make the whole soldier, we must not forget the **stands themselves** for the roads.

If a strip without trees is to be left beside a road, as demanded so often with improved roads, then some people think it is beautiful if the neighboring stand presents itself as a wall of leaves, everywhere closed down to the bottom. Already Koenig set up the demand, "to maintain **all** the edges of the forests as much as possible closed and green." It has already been proven in this book that this demand in general goes too far. It is nice, and it is interesting to **see** every now and then **into the interior of the stands,** also to be able to admire a trunk under a more favorable **angle of sight** than would be possible in the interior of the forest. In France, beside every chaussee [high road] a very wide strip of land is kept free from woods in order to make the chaussee dry out more easily. Beyond this strip, the view towards the forest meets everywhere a monotonous, long, hanging forest margin. After a time this becomes boring to desperation. A higher officer, who had been riding from Woerth to Hagenau, that means through the middle of the magnificent Hagenau Forest, could complain to me with reason that he had not seen even one large tree the whole way. The richest forest of the Empire had appeared poor to him, because it had not really presented its treasures.

The tree rows on both sides of a road are not always planted in a parallel way. In order to make an allee appear longer than it is, one makes its lines come closer to each other in the distance. One of the most well known examples of this kind is the Bellevue Allee in the Berlin Zoo, but it is laid out so wide and closed down so much, that the intended deception does not succeed.

If I have been reproached several times for wishing to form the forest into a "pure park," then the length of this chapter will prove how inappropriate this reproach is; for it is known that allees do not belong in a park. Even Petzold, although otherwise an emphatic friend of allees, admits in his *Landscape Gardening:* "But totally inadmissible is an allee in a landscape garden layout of modern style." If I now recommend the establishment of some allees in the park-like landscape, and many allees in the forest, this is not meant to make landscape and forest into parts of the park, but rather to distinguish them quite clearly from the park. However, an especially diligent care for the allees will be a mark of those "enhanced forests" with which this whole last part (II B) of *Forest Aesthetics* deals.

While there is a sharply distinguished difference between the field of forestry art and the one of **modern** landscape gardening, a clear separation line is missing between forestry art and this **older style of gardening,** whose characteristics lay mainly in the layout of straight-lined shaded walks.

For this reason it is a failure of the **Zoo of Berlin** that in some places one cannot tell there whether one is still in the enhanced forest, or, like the name already hints, in the (animal) **garden.** On the other hand in the Eilenriede, the city forest of Hannover, it has been done very well, in doing without straight-lined allees and any other splendor (for example foreign tree species, etc.), so that the forest parts were set into a real contrast to the parks and gardens of new and old style, which are so abundant near Hannover.

Chapter Six

Old Trees as Adornments of the Forests.

Burckhardt says: **"The most beautiful, however, that the forest has are its venerable trees and stands." However, "the old tree stand has to fall at last, but save it where it is a rare appearance, until other considerations will demand their right. But the old hermit tree, the witness of the mighty power of nature, which has**

been passed by centuries and whole generations with their history, which maybe carries its special name among millions of trees and which, widely known, has seen quite a lot of sons of the forest under its roof, who have been dead now for a long time—grant its place to it, until it is broken by the storm or its last leaf is bleached. Then put a young tree there in its place, as a memory and as an heir of the name, a mark of the place in the wide forest."

Such respectful action is also beneficial in aspects other than the aesthetic one. Old trees are a very valuable illustration for the taxation notebook. In his strong words, Koenig expresses this idea in the following:

"Rare, particularly large magnificent trees and stands should be maintained as long as possible, even if ordinarily grown trees would need to stay, too, for their support. If we totally destroy the last giant remnants of prehistoric times, then nothing will remain that might be able to remind the future **of more devoted obeying of eternal natural laws;** the tiresome egoism might at last think of the artificial dwarf figures of the new forests as being justified and all right."

In respect to the oak, we generally must not complain, for rather too many than not enough are saved as standards; at least it is like this here in Silesia; but very rarely other tree species are allowed to mature up to a venerable age. Foresters and forest owners probably are not guided, in doing so, by a conscious judgement of taste of their own, but preferably by the continuing influence which poets and painters, by word and painting, and the public opinion dependent on them, have on all of us. There are even whole districts where not even a single older oak is cut, may it be declining as it wants to. Even if there is nothing more than bald remnants, as long as it can still present a bundle of green leaves, it is left to stand. Even where they have little effect by their position in the district, even when there is still a rich abundance of old oaks, they still enjoy the preferred treatment to be saved from the axe by any means. Obviously this goes too far. **Retaining too many sick trees is no adornment for a forest district.** It would be barbarous if someone would want to put the axe to the remnants of a six-foot strong oak, if he possesses only one or a few of them, but rotten trunks numbering in the hundreds are no embellishment for the forest; therefore we have to join the mycologist if he demands to reduce the number of breeding grounds for fungi spores, which cause disease.

So, for readers trained in forest aesthetics, R. Hartig would not have needed to defend his point of view against misinterpretations, as he did in a remarkable section of his *Signs of Decay,* for three generations earlier than this, Gilpin had already arrived at the same point of view when he complained about the oaks of New Forest: "Many of them are damaged and scrubby. When being part of a combination, trees like these may be able to have some effect, but if they appear too often in a rich forest scene, then they will offend the eye."

When Gilpin writes "a combination," he means that the oaks, beginning to decay, have to unite with other younger woody species, with a building or a rock, into a complete scene, but that they must not stand **individually in a wide open area.** In such a position even a healthy tree has a melancholy effect (one might remember Heine's: "A spruce tree is standing alone"), and the gloomy impression caused by a dying tree would be too strong. However, the sublime carries high intrinsic beauty even in declining—**the tragic beauty;** but if we want to realize such beauty, we must not neglect how the poets begin: they do not make their **hero** stand all by himself; they surround him with mediocre characters who support or fight him, but no one of them is able to measure up to him. So we, too, should not let mediocre trees decay, but let decay only very mighty trunks (old warriors, who have gone through quite a few storms) as remnants. If we did it another way, if we wanted to save, besides the very old gigantic trunks (be it just **one,** or be it several ones, united into one group) the equally strong but still a bit inferior trunks, so we would lose the right scale for their height, or, to say it better, we would gain a scale whose application would be disadvantageous; for the strongest ones are easily compared to the less strong, and these with the weaker ones, and so one ends up and thinks even of the first ones of not being overly large anymore. But whatever we can judge and estimate ceases to cause the impression of the sublime for us, and the tragic impression can diminish until it becomes merely miserable. Relationships with great mass do not always enjoy an advantage; often it is the more advantageous position that becomes decisive for choosing which of the trunks are to be saved. I remember a cutting area on which about twelve punky oaks had been saved, first as a trial, in totally irregular distribution; but soon it was discovered that, scattered as they were, they did not enhance the district in any way. Quite the contrary, they gave the impression of wastefulness. That is why only those three that stood at the most advantageous place were maintained. Now it is easy for the eye to unite them into a group, and they appear more imposing than the unordered load before.

Remarkable or particularly beautiful old trees not only have to be saved from the axe; one also should take actions for the **extension of their lifespan.** I hope to please the reader, and I am very thankful to Fickert for this information, by reporting in his own words of the notable way in which the Hertha Beech on the island of Ruegen has been maintained until today. Fickert was so kind to write to me:

"When I came as a chief forester to Ruegen on July 1, 1852, I found the Hertha Beech, which I had known for 8 years (I made a management plan for the forest ranger Werder in 1844/45), in a sad condition. It was carrying hardly a quarter of the leaves that were supposed to be there, and the existing leaves were hardly green, but gray and yellow and half of the size they were supposed to be. I was very sorry for that; **this one** now also goes the way of all flesh, but will leave a very perceptible gap on earth—so I thought—but I did not think of the possibility to help. Then help came from above. In August, His Majesty Friedrich Wilhelm IV came to Stubbenkammer. Also, a walk to Hertha Lake was undertaken, at which occasion I was walking far behind in the long row of the retinue, with Count Stolberg-Wernigerode, the minister of the Royal House, when the Landrat [head of the administration of a Landkreis] arrived out of breath: "Mr. Chief Forester, His Majesty orders you to come to him!" Well, I got going. "But tell me, the beautiful Hertha Beech is dying after all." Yes, Sir. "Yes, Sir, what the heck, I have not given any order; but tell me, what is this?" Probably it has reached the end of its natural lifespan, Sir. "Of course not, natural end; that cannot be true, that must not be true; oh, if I had known that then I would not have gone here, my beautiful youth reminiscence!—but listen, dear Chief Forester, treat it rather nicely now for a while, help it to get on it's feet again; but listen, really do it!" Yes, Sir. "Yes, Sir, he says," the King remarked, turning towards Hertha Lake again, "he will probably also not be able to help it." I took this event to heart and I immediately started the investigation. Count Stolberg said: "Well, dear Chief Forester, this is a bad task; you have to do something, for if His Majesty, as expected, will return next year, and then you might not be able to tell and show him what happened, then His Majesty would be very ungracious; let us take a closer look at the turf bench." This was done, and we soon discovered that the turf bench consisted entirely of roots, upon which a lawn cover, in a somehow miserable existence, was lying. So we had already discovered essential things, and I had to blame myself for not having thought myself earlier of the idea to try to save the tree. In the evening, His Majesty talked about the beech again and said: "Now, don't forget about the Hertha Beech", and when I could answer: Sir, I think I have already found a notable hint, the King was visibly delighted, and I had to tell him about my project. Then I visited the oldest forest worker, a man of 82 years, and had already found in him the only one who was still able to give me the most useful information. Eight days later he was dead. He testified: About 1822, I got the order from Chief Forester Koehn, who had been brought over from Sweden in 1815, to dig away the little hillock on which the beech was standing, to level the place carefully, and to leave enough standing around the beech so that a turf bench remained, like it is still today. Many roots were cut off, so the operation seemed to me to be very dangerous for the beech. The Chief Forester's wife was a very merry woman who loved to dance and wanted to dance around the Hertha Beech; therefore this was done, and later on music was often played for dancing there.

Now I irrigated the turf bench heartily, until all the roots were free [of soil]; it was quite a dense matting of roots; like a dense brush the fibrous roots were surrounding the tree in the shape of a garland. An embankment of good humus soil was built around the bench, and this material was filled together with water between the roots, until the roots and the space between them and the embankment was completely filled; it was worked on like this for five days. Then the terrain around the beech was reconstructed and this was done with good humus soil, like old Arndt had described to me how it had been before. The next year, the foliage looked much better and in 1854, it was back to its old abundance. In August of that year, the King returned, but unfortunately Count Stolberg did not; he had already died in the meantime. First I reported to our minister, Excellency von Bodelschwingh, and I told him of the surprise that was waiting for His Majesty, and the previous story up to the present. His Excellency, very delighted, said, come, you have to tell that to the Prime Minister, Excellency von Manteuffel; this was done, and soon the whole retinue was in the most joyful mood for the fortunate success; but be silent, do not tell! Then we were ordered to the dinner table. As His Majesty saw me, he threatened with his finger in a friendly manner and asked: "The Hertha Beech?" Good, Sir, I answered. "Your luck," His Majesty said, "first eat and drink and then look at it." His Majesty was very cheerful. All of a sudden, towards the end of the meal, there was a general movement; His Majesty was carried away, brought to bed, and brought on the ship the next morning; he did not return to Ruegen, for soon after that, the incurable disease broke out, and so he did not enjoy the pleasure of seeing the notable tree again in its rejuvenation. I have enjoyed it for 22 more years."—So much for Fickert.

The Hertha Beech has a circumference of 4 m close under the beginning of the crown, which takes place at a height of 1.3 m. Its age may be 470 years. (The former Chief Forester Smalian had bored into the trunk in 1844 and calculated the age, it being 410 years at that time). Even today its growth does not leave anything to be desired.

Only in individual cases the situations may be as difficult as they were with the Hertha Beech. For my strongest oak here in Postel, it was enough to cautiously loosen the soil below its crown and to add several cartloads of leaf mold in order to inspire it to a new powerful growth, after it had been ailing for quite a few years.

This "thick oak" has a circumference of 627 cm at breast height; its diameters are 198 cm and 204 cm. There are 10 annual rings on the outermost centimeter, and I estimate its age to be about 400 years.

Much more important is the Klipstein Oak, standing near Darmstadt. Its circumference at breast height is less (6 m), and the diameters are 1.8 and 2.3 m. Still at a height of 5 m it has a diameter of 1.9 m, and the mighty crown starts at 9.5 m.

Wilbrand tells about a successful operation, undertaken for the saving of the tree.

"Wasps were seen to be flying out of a highly-set, strong, rather steeply rising branch of the Klipstein Oak. The tree was climbed and examined. It turned out that at the spot where two mighty branches were touching each other, a hole had developed in the lower branch and extended far in the direction of the trunk. It was thought to be advisable to remove the sick branch. A roofer took over the job. Starting at the tips of the twigs, parts of the lower sick branch were attached by a rope to the upper healthy branch, and were sawed through in short sections, each behind the attached ropes. The severed pieces were let down with the rope, in order that they would not cause any damage in the understory or on the gravestone monument of President von Klipstein, which was situated below the oak. So the strong branch, which offered more than 5 cubic meters of fire wood, was cut off up to the trunk. It turned out that the rotten part extended only a little way into the actual trunk. This was carefully scraped out and filled with asphalt; then the cut surface was thickly covered with a mixture of asphalt and cement, and was shaped and marked to appear like outer bark. The tree was saved, and few of its numerous visitors notice the immense operation it has gone through."

In the Granducal State Forests of Darmstadt, much has been done for the saving of venerable trees, even with rot already advanced. The tried and tested instruction, which I got from Wilbrand, is quoted here:

"Before filling a hollow tree, be particular that all the rotten wood on the side walls and especially on the bottom of the cavity is carefully removed. The laying-up with bricks (rough stones will not hold for very long) has to be done in a way that a wide foundation (rough stones may be used here, if at all), which fills the bottom of the cavity, is created; upon this a narrow layer of bricks is lined up, the upper part gradually narrowing, but it has to fill the cavity at the bottom, but does not need to fill the whole cavity. This lining-up however, can be done so strongly, if necessary, that it can serve as a holding and support structure for the tree at the same time.

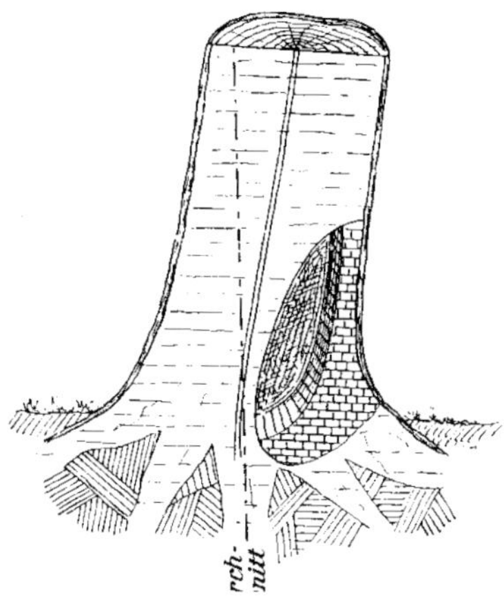

Fig. 55

"For laying the bricks, a mixture of 1/3 of cement and 2/3 of sand (no lime) should be used. For filling the gaps at the sides, stone chocks are used, and finally, any remaining hollow spaces are filled with cement.

"For the outermost cover, half cement and half sand are used; for the creation of the bark-color some gray, green, maybe also yellow color is added. The shape of the bark is re-created with a delicate trowel.

"The work should be given only to a bricklayer who is experienced in working with cement."

Now I have stayed very long, nearly too long, with oak and beech, and now I have to be brief with the other tree species. For the maintenance and growing of handsome trees, Koenig lists as a reason that these are to remind posterity of the **"more faithful obeying of eternal laws of nature."** Our actions correspond with this purpose only when we strive to tend **all** the tree species which are native to the site or, being extirpated by our actions, once were native, at least by some handsome fine specimen of each species, in order to make the "eternal law of nature" that rules in the vegetation visible in as many facets as possible. The venerable, now nearly extinct **yew trees** owe the maintenance of sparse remnants in some districts only to such respect. They are standing there strangely, in the new era of clearcutting, without descendants, like Chingachgook, the last Mohican, dreaming away a solitary existence, mostly beautiful in their way, however always interesting.

Unfortunately, it is a fact that not every forest owner can provide himself soon with a thousand-year-old *Taxus,* but even a large poplar, an old birch in free position, a strong alder, are things very beautiful in their own way, and one generation is enough for these in order to let them grow from common size into relatively extraordinary strength. In the yearbook (for 1883) of our Forest Society, I found the obituary that Kirchner dedicates to an alder of his district. Under the title "Grieving Things from the Forest," he reports its death. With a diameter of 1 m, it had a height of 32 m on the clear section of the trunk. It had become only the victim of an "unfortunate misunderstanding." This example shows what can become even out of alders; one only has to let them grow, and where the site does not support ones that large, even smaller ones can make an impression.

If it had had a name, the alder would probably not have been cut down. I think that giving a name is a very appropriate way to secure respect and saving for notable trees. To have a "Danckelmann Pine" cut is something that a superior will not easily stipulate to the chief forester.

Also, the entry into the taxation note books (in the chapter describing the forest district) and into the forest botanical notebooks is appropriate in order to secure the maintenance of notable trees.

The treasures that exist within a district not only need to be taken care of. They also need to be presented in an advantageous way.

The sites of old trees need to be made accessible, but it would be a mistake to make each single tree giant be seen in quite the same way as all the other ones.

For some of them, a road can be located close by. The majority should be shown only from some distance, sometimes a greater and sometimes a lesser distance. Not all of them are to be released totally. With some of them, the trunk should be made visible, while with others, only the sight of the crown should be opened; in doing so one should consider the special advantages of each tree.

I have opened views from three directions onto the strongest beech in Postel, the Emily Beech. Plate XV shows such a view, released last year.

The measurements of the Emily Beech are: circumference at breast height 420 cm, diameters 135 and 145 cm, height to the base of the crown 3.50 m. I estimate its age to be 250 years.

Chapter Seven

Aesthetic Use of Foreign Tree Species and of Varieties of Native Tree Species

The attitude is widespread that the use of foreign tree species is a major means for enhancement of the forest. I cannot share this opinion, and I wish to try to prove that we can get along very well with the treasures of the native flora.

"The bread of someone else is cake to the children." This Silesian saying has been true for me, too, inasmuch I overestimated the value of foreign tree species in my forest-aesthetic childhood. There is even less excuse for this, as many failures were standing in front of my eyes.

My father has planted horse chestnuts and locusts in a great number in the forest, and he has a very large field planted with white alder. Of the horse chestnuts, favored by being released, 15 are able to maintain themselves. Of the locust, only two are still participating in the stand. The white alders have stopped growing. They have been replaced mostly by vigorously growing oaks, which owe their existence to the busy jaybird.

I myself have planted white pine and larch, also noble fir, and particularly numerous spruce, which are not native on our sites. In Postel, the deer have destroyed my white pines by peeling and abrading, while in Klein-Commerowe, most of them have become victims of a disease, the Blasenrost [blister rust]. In the beginning the larches were growing very well, but last year they all (except for a few remnants) have become victims of a fungus. The noble firs, about 30 years old, have not grown yet beyond my own height. One almost needs a magnifying glass to find many of them. Even the spruce does not keep what it promised. Being invaluable as a filler of gaps, it disappears from the developing stands. It is barely able to maintain itself between the oaks; standing among beeches, it very soon gets overtopped.—Whenever a plague of non-moths [*Lymantria monacha*] approaches my boundaries, I spend hours of worrying because of my spruces.

Even if I, too, do not shrink from further tests, and recently have undertaken extensive experiments with Sitka Spruce [*Picea sitchensis*], Douglas-fir [*Pseudotsuga menziesii*], Black Walnut [*Juglans nigra*] and American Ash [*Fraxinus americana*], then nobody can regard me as a fundamental opponent of foreign tree species.

But I grow these tree species as an experiment with the hope that they will serve well for forestry purposes, not for the embellishment of the forest, and therefore I hide them in the interior of the stands.

This method has found the support of Hampel, who said the following: "The custom of completing places which are extremely eye-catching with non-native tree species, which usually do not harmonize with the native tree species, or mixing or framing them with foreign species, has to be rejected according to my opinion; besides the fact we do not yet have enough experience with foreign tree species, this custom has to be called unsightly; experiments like these actually belong somewhere else."

But much more numerous are the opposite voices. As the remarks of Mayr and Hartig appeared especially notable to me, they will be discussed later on, after I have explained my own point of view. I recommend the use of foreign tree species:

1. In the **experimental forest,** specially dedicated to this purpose, as well as on little **experimental areas, hidden** in the interior of the stands.

2. Everywhere where it has been proved by long years of experience that they bring better results in respect to forestry or hunting than native tree species under the same circumstances (game preserve, wildlife food plots!)

3. In reforestation of waste lands.

4. In small forests, where it is important to create as much variety as possible on a small area.

5. Near foresters' houses, so to say as a sign of cultural industriousness.

6. On improved roads as allee trees. The latter types of use guarantee the most freedom, because the strictly forestry purpose—the growing of the trees for their revenues—has less weight compared with the considerations of other purposes.

If I cannot say more towards the supporters of foreign tree species, my reserve is partly based on personal taste; for I think that the

XV.
Postel. Emilen-Buche.

foreign tree species do not really harmonize with our tree species, and I am not always able to give a detailed reason for this unfavorable judgment. But this may be disputed; other people with other tastes may judge in a different way; but it is certain that mistakes in the plantation can easily be made with exotic tree species, not yet sufficiently tested, which will increase the unfavorable impression. If a foreign tree species that attracts the eye of the passerby is sickly, then this does much more damage than the same phenomenon with native species.

However, if we give only the best sites of the district to the exotics, prepare the soil by deep loosening or even compost, then they will develop very rapid growth in their youth, and the native tree species stand beside them like Cinderellas. Particularly, the modest adornment which our forest is going to put on in fall, touching eye and mind so pleasantly by soft shading of the colors, is actually crushed to death by the garish colors of the exotics. Mistakes in the selection of the plantation method and of the combination, however, will decrease with increasing experience. They may be avoided altogether; even the most skillful application faces two objections:

The foreign tree species **disturb us in the illusion to be "in the open air," to be surrounded by nature left to itself, and they decrease the desired contrast between forest and garden.** Namely, the garden is the right place for foreign species. I found this justified in detail by Bratranek, who writes on the garden plants: "The more different from the commonplace, the more they display the far-reaching influence of the owner. . . . Because the garden is a place where the human enjoys the daily satisfaction of needs towards his human part, it is understandable that he also tends only the extraordinary plants in his garden as flowers. Even if the native plants growing around him show very nice attractive forms and colors, he will not take them into his garden, if only because the garden is his planting, and that which presents itself everywhere voluntarily reminds him too much of wilderness, which he strives to overcome by his cultivation."

Probably a similar thought has caused the Baron von Hammerstein, being the Prussian Minister for Agriculture, State Estates and Forests, to give the order that the surroundings of the **forest official's dwellings** should be shaped in a more inviting manner by planting tree groups with the particular use of foreign tree species, like the Douglas-fir, etc. The expenses increased by these actions are to be paid out of the planting fund.

This would be quite all right with me, if the Douglas-fir had not been named in the first place. The forester likes to have trees in his small garden, which are beautiful and useful at the same time. One should plant for himself fruit trees, preferably species which are not only usable but also beautiful, like walnut trees and peach trees with original roots [not grafted]; one should provide bee pasture like the American red maple offers already in April, like common locust [*Robinia pseudoacacia*] and sticky locust [*Robinia viscosa*] and American linden offer. Where the soil and the climate allow it, one should use *Sorbus domestica,* which is as beautiful as it is good for the preparation of fruit wine.

But I cannot go beyond these hints here. The establishment of gardens at foresters' houses may be a chapter of its own in a future edition of *Forest Aesthetics.*

If one acknowledges with Bratranek and Baron von Hammerstein that foreign trees are to be planted in the immediate vicinity of the house, then one must also admit that they do not belong in the interior of adjacent forests. When we "escape" into the forest, then we want to forget, for some moments at least, that it has not been the motherly care of nature **alone** that has worked here. But how can we maintain such illusions if we meet foreign trees wherever we go, which never would have been able to intrude here or to even dominate without human intervention, when we meet the same old acquaintances again in the forest, that hit us, well-groomed, right in the face with every look we take out of the window into the garden.

Now, the latter objection is not true everywhere; for not everybody possesses numerous foreign tree species in his garden. Also,

I have to admit that in some exotic trees there is indeed inherent a certain value for use, due to which we can not banish them completely from the forest. In northern Germany this is true for the larch, the noble fir, the white alder, in all of Germany for the white pine, the horse chestnut, the locust, but especially for the Canadian poplar. Respecting these facts, however, I stay with the opinion that in **large unbroken forests patches of exotics must not be planted around everywhere in so-called experiments for acclimatization; but in very small conditions, when the forest land area is insufficient to give an impression of a forest, and the soil is too poor so it cannot substitute for the insufficient size of the area by lush growth of large trees, then it could not be further spoiled by planting exotics. In small conditions like these, I would say even petty conditions, delicateness, neatness and a certain adornment will most of the time have the very best effect. Under these circumstances, the cautious use of Weymouth pines, black pines, red oaks, locust, etc. is not only allowed, but even advisable for embellishment, mainly along the roads.** The locust can be used very easily as a **nice understory** even on poorest pine soils. It is necessary to plant only a few little stems of locust together with the plantation. With each thinning, some of these can be cut down, in order to inspire root sprouts, which stay alive for many years below pines, as experience shows.

But, **in large forests,** it has to be proceeded in quite a different way, as has been said. Someone who wishes to establish American and Asian trees in those, be it for dendrological passion, be it with the hope that they will be useful and pleasing for his grandson, should separate a suitable piece of land as an **experimental forest,** should banish anything native there and might establish there a piece of Canada or California or Japan, and he should also name it as such. It will hardly turn out to be a gold mine, but it may be beautiful at least. White pines and Douglas-firs, the red oaks, the silver maple, the black walnut, along with many ashes and birches, form the stands and are used as allee trees; but at the edges and as understory, hemlock [*Tsuga canadensis*], white spruce [*Picea glauca*], eastern redcedar [*Juniperus virginiana*]. A wealth of magnificently flowering shrubs might complete the ideal. I might think that somebody who owns 1,000 ha of forest and additional wealth, and who enjoys things like this, might do well to give about 100 ha to such experiments.

I had the opportunity to admire such an experimental forest at the home of Count von Wilamowitz-Moellendorf in Gadow.

Where it is the objective **to establish a new forest in a treeless region,** there the forest establishment in itself can be regarded as a daring attempt, and the whole forest will become an experimental forest just by itself. Then it will be quite **stylish** (to use a modern expression) that this forest also will give evidence continuously and clearly of its character as an experimental forest by the tree species in it. The wonderful firs of Luetetsburg, whose high dark crowns I saw defying the North Sea storms on the East Friesian coast, certainly will not be astonished, being themselves foreign in the region, that also Douglas-firs and other exotics will wander beyond the park fence up to them.

The remarks above have been taken with slight corrections from the first edition of Forest Aesthetics, although very influential persons do not agree with my view. For example, among others, R. Hartig writes:

"I do not mind admitting that we do not **need** the exotic trees in the deciduous forest there, where we deal with favorable site conditions, in order to achieve the highest goals of forest aesthetics, and that there a stranger **may** even **disturb** us **under certain circumstances.** But in this, one will need to distinguish among the individual species. Douglas-fir and Nordmann fir as a rule should not disturb, because they hardly cause a foreign impression with us, but I do not mind admitting that cypresses might disturb us in the forest. Due to their strange disposition, they can diminish certain illusions."

"On the other hand I have to speak for the use of exotic trees in the most monotonous forests, particularly conifer forests. When the eye and the mind are tired while walking through monotonous pine or spruce forests, a single tree or a group of beautiful exotic conifers have an effect on the forester's mind like the sight of a good roe-buck."

"Like we decorate our houses with paintings, with statues, with beautiful plants, in order to make the stay in them pleasant, we also can bring a welcome change into the monotony of some forest areas by the planting of beautiful exotics. After all, our modern conifer forests, particularly in the plain, are a long way from creating the impression that they 'sprang from nature left to itself.' Within them, a group of beautiful exotics as a rule will not be able to destroy any 'illusions.' I do not doubt that Mr. von Salisch will admit to the exotics their legitimacy in the German forest, at least within this limitation."

Now, that much is surely true, that in some monotonous forests nothing could get worse. One will welcome any change within those, and nobody will think of an actual beautiful tree species, like Douglas-fir or Nordmann fir, as unsightly, if they thrive. But do we need them?

If we dedicated the same expense and the same effort to embedding decidous trees, we shall achieve by beeches, oaks, birches, maples, the same, even more, and in this we avoid the risk of unsightly failures.—The overtopped beech is welcomed everywhere as a valuable soil protection tree; but miserably growing foreign trees are looked at with pitying eyes.

The kind reader might want to take the effort to go back to Chapter Four, Part I, B, in order to check again the great number of usable varieties of native tree species; one probably will have to admit that for every aesthetic purpose the suitable material presents itself within the native flora.

I would like to finish this discussion with the reference to Prince Pueckler, who expressed himself with the following, concerning the issue of exotic trees:

"As a rule, I use only native or totally acclimatized trees and shrubs in the park, and I avoid completely all foreign decorative plants there; for also the idealized nature still has to carry the character of the land and the climate where the established area is situated, in order that it will be able to appear as having grown up by itself and will not reveal the effort that has been used upon it. We have quite a lot of flowering, very beautiful shrubs which grow wild here with us in Germany, and they might be used at many occasions; but when one finds a *Rosa centifolia,* a Chinese lilac, or clumps of such shrubs in the wilderness, this causes a very disgusting, affected effect, unless they are in a separate space, enclosed in itself, for example a small fenced garden beside a cabin, which in itself sufficiently signals the closeness and culture of humans. Some foreign trees, like the white pine, locust, larches, sycamores [*Platanus*], honey locust [*Gleditsia*], and blood beeches, probably can be taken for completely native; however, I still prefer with us the linden, oaks, maples, beeches, alders, elms, horse-chestnuts, ashes, birches, etc."

All the reasons explained by Pueckler are valid in the forest even more than in the park.

Some particularly conspicuous **varieties of the native trees,** like for example the ubiquitous blood beeches, weeping ashes, and strangely foliated oaks, are standing stranger still against their original forms, because of their appearance, than some of the foreign members of their genus. Also, their juvenile growth is mostly so weak that they do not overcome the diseases of youth if not on cultivated soil. **Therefore they will be touched by the objections explained at the beginning of this chapter, but on the other hand it is in their favor that their use is not absolutely against nature.** One might even say a tree species, for example the oak, had not yet gotten its rights in a district, as long as it is not represented with all of its characteristic varieties in this district. Those who manage in districts which have for generations been subject to the all-leveling system of clear-cutting, like it is unfortunately the rule nowadays, will rarely be sufficiently aware of how richly and often nature likes to extend the forms of a species into different directions and to alter them. In the close vicinity of my place of residence—fortunately, here the conditions are not that much civilized yet—I accidentally discovered without looking for them, sessile oak and hornbeam of beautiful pyramidal growth, beech and spruce with drooping branches, and a *Picea abies* 'Virgata', as well as two different dwarf spruces (similar to the well-known garden varieties '*Gregoriana*' and '*Clanbrasiliana*'), and oaks and birches with foliage conspicuous of form and color. These are varieties so clearly characterized that 40 years ago, when the assortments at the nurseries was not yet so abundantly available, they would have belonged to the most valuable novelties. **It is surely advisable to maintain, by special care, in a district such gifts offered by nature itself.** Also it seems to be quite safe to take some grafts from the mother tree for greater security, and to grow **descendants** from it **in the forest nursery** for occasional use; and why should the seed not be used that nature produces of its own will? Particularly the varieties most distant from the original form often prove to have very characteristic seed. In the Kreis Trebnitz (Koschnoewe near Praussnitz) there is an oak whose crown is as yellow on its sunnyside as *Quercus ped.* '*Concordia*'. Being about 100 years old, it bears rich seed, and this seed produces strong seedlings that partly carry even lighter-colored yellow foliage than the mother tree. (Such seedlings came to the nursery Monhaupt in Breslau, and I think it is not totally unlikely that the Concordia Oak, so widespread nowadays, stems from there, with a detour through France.) Now, who can blame the owner of this oak, who can blame his neighbor, if they wanted to choose preferably yellow oaks as an allee tree, using the existing abundant seeds? Nature itself would have given the hint for that. It is known that the blood beech, the maple whose leaves are red on the undersurface, the pyramid oak pass on their characteristics almost surely by seeds. They are at least—certainly the beech and the oak—of German origin, if, however, not native to Silesia like this golden oak; but why should we restrict ourselves, why should we, in particularistic narrow-mindedness, do without and not plant an effective alternating allee, for example, out of blood beeches and golden oaks? Now, it is certainly not particularistic narrow-mindedness that will prevent us from doing so, but more a correct understanding of nature. **Nature everywhere puts a wise restriction on itself, it does not fish for cheap effects. Only rarely and individually it produces these "natural plays," and so we should only use them rarely and individually.** Whoever might wish to have an arboretum outside the garden, out of dendrological passion, might combine assortments along the road as an **"open allee."** This is the only way that a planting of great numbers of this type will be allowed to be done outside the garden without worrying. There are minor objections towards the abundant use of only **one** variety, and here it is especially the **Pyramid Oak** with which I would be willing to make considerable concessions. At least this one is justified, being native, to replace the Pyramid Poplar, over which it has such great advantage. It was discussed previously what sort of role it has as an allee tree. **Pyramid trees must not be used individually in the landscape, but only in groups of at least three, unless they are standing at gates or bridges as a sort of mighty pillar.** They fit very well at places like that, as well as everywhere where it is important to point something out from the distance (like the site of the home we strive towards, or the crossroads we must not miss). They also possess the advantage that they offer a good scale for relations of height. Being put at the bottom of the basin or at the edge of the hill, they make the relief and sinks or bottoms appear larger than they really are. Subconsciously we overestimate the height of their extended forms, because they are so important in relation to their small width, and then the error favors the appreciation of the terrain conditions. This happens just like one will always sketch a top hat, hanging on the wall, about one third too long.

Where it is the goal to interrupt long horizontal lines, there pyramid trees are also at the right place. Therefore we like to put them in front of buildings and in front of an extended horizon, or we make them tower above a wood wall, which limits the view in a straight line or in monotonous wavy lines. That is why they also fit very well at the water's edge. A composite forest which accompanies the banks of a stream will often offer an opportunity to use Pyramid Oaks in the described way, and whoever wants to have an effect sooner, might want to plant Pyramid Poplars among them. The picture XVI may help to explain the things said above.

For all the trees whose appearance differs considerably from the surrounding, be it exotic or be it native trees, the rule brought up by Prince Pueckler is valid: that they should not be used on only one side along roads, but that one should consider a certain equilibrium.

This is true not only for individual trees and groups in the park-like landscape, but also for the interior of stands.

If one happens to find—I take this example from the Beech-bank-line in Postel—a pine thicket, irregularly bordered by spruces on one side of the firebreak, and 70-year-old pines on the other side, then it would be wrong, to tend below the latter only the natural beech and oak seedlings as an understory; at least at the edge some spruces have to be added.

Chapter Eight

Embellishment of the Forest Stands by Care of the Shrubs and the Groundcover

The grazing rights of former times, the litter removal and the system of clear-cutting have damaged our shrub vegetation so much that in some districts not even half the species are left which the area could potentially contain. This ought not to be like this, and it ought not to remain like this; for **how beautiful are our shrubs, how different are their species, and how manifold changing is the appearance of each species,** according to whether it was allowed to grow in a free position or at the stand's edge or under

XVI
Rhein-Landschaft.
Blick von der Insel Mönchsau auf Reichartshausen.

the more or less sparse canopy of the pole stands and the mature stands, how they decorate the forest by leaves, blossoms and fruits, and how much they **indirectly decorate the district by the wildlife which they support with food and cover.** The king of the forest, the **noble red deer,** is miserably dependent on welfare at the game feeding station where heather and juniper no longer offer even the most modest winter browsing. Not like a king receiving due tribute, but like a beggar, it receives the measured ration of shriveled horse chestnuts. Lesser inhabitants suffer even more, particularly the **feathered songsters.** How poor is the bird's world, according to numbers and varieties, where it no longer finds any more food and cover within the shrubs, even through the fall! Even the **insect's life** gets poor. Where there are no shrubs, the **butterfly** is no longer an adornment for the district, but a danger. Where the caterpillar of a Schiller butterfly can no longer live on a willow, where the brimstone butterfly looks in vain for its Schiessbeere leaf on which to put its eggs, is just where the unwelcome guests (the clumsy silkworm moths and the flapping geometer moths) like to come the most.

The majority of my esteemed professional colleagues are so far from caring for the shrub vegetation that I can hardly hope to have convinced them to consider shrubs by the listing of our plant treasures in Chapter Four, Part I, B.

Other people have already recommended in vain the artificial growth and cultivation of decorative shrubs; Burckhardt, for example, has supported that. In the essay, published only after his death—*The Nurse Tree Species and the Way They Work*—he lists several small shrubs as being desirable. Some, among them two foreign genera (*Diervilla* and *Mahonia*), he wants to be grown partly from seeds, partly by root suckers.

The saving and maintenance of existing shrubs is easier than new establishment. If it happens that on a plantation area intended for planting spruce a shrub has also become established which is rare in the forest district (out of my own experience I think of *Daphne,* red-berry elder and barberry), there it often takes only a minor effort to put it onto an appropriate place in the neighborhood, where it will be able to develop further without disturbance. Also during the weedings and thinnings, quite a few things still can be saved. What gets awkward on one site, occuring in great numbers, may deserve care as a rarity some miles away, for example the **ivy**. In my deciduous stands, it eked out only a miserable existence, creeping along on the ground in long tendrils. Its attempts to climb upwards always failed, and I thought this was due to the winter's frost, until I discovered that in winter the deer tear each tendril when browsing the leaves off. Soon I helped this problem with some hawthornes, and now I see with pleasure the ivy rising up high on several trunks.

This is the most difficult in wildlife preserves and fenced game areas, while, at the same time, these very places, being "gardens" in a certain sense, are supposed to appear particularly beautiful.

Whoever takes a look into the wild pig area in the Grunewald and compares the pleasing lushness of the shrub vegetation with the pine stands of the rest of the district, being totally bald and barren at the bottom, will complain about the raising of a too large fallow buck population and will wish that this wildlife species would be replaced partly by black boar.

In the Schorf Heath, I saw hawthorne, barberry, wild gooseberry and juniper maintain themselves, despite the large deer population.

To the kindness of the Castle Captain of *Heimburg* I owe the information, also proved elsewhere, that the Pontic Rhododendron are spared in any case even by the fallow bucks. It is a pity that this marvelous wood has certain needs regarding the site, and that it can survive the winter frosts without damage only in preferred parts of Germany.

There are quite a few districts whose peculiar character is determined less by the woody species than by the **non-woody plants growing on the site.** Those bottomland forests, whose ground is white with **snowdrops** in early spring, then yellow with **cowslips,** those foothills where the **liverwort** seems to reflect the blue spring sky from the ground, also those who scent the air, covered by the dense foliage of the **lily of the valley** (we Silesians call it Springauf), out of a thousand white cups, **they all make an impression on our mind, more due to the flower decoration than to the tree species they carry.** Therefore it suggests itself to try to establish such welcome adornment where it is missing, but unfortunately I have to admit that this is not as easy as one might think. Those children of the spring are very selective in respect to the site, and only very little success can be seen from my efforts with them over many years. Maybe other people are more successful. So I saw at least **one** successful attempt on the sandy soil of the Mark (partly of Class II, partly Class III for pines) in Rogaesen near the Royal Forest Administration of Altenplathow. There strawberries, ferns of different species, lily of the valley and the like, have been planted into a pine stand established on an old field, an embellished grove close to the garden, just during the first thinning, with such a success that original, uncultivated forest ground could hardly be richer with them. Due to some soil preparation on appropriate moist sites, now even *Digitalis* thrives there in surprising lushness and regenerates naturally. Being a souvenir of a mountain journey, the plant had been added as the only exotic to the other plants, which were all characteristic for the site.

However, such experiments succeed only as a real exception (due to the failures of many years, I unfortunately have to repeat this.) But the more difficult it is now to establish missing species, the more eager the forest owner should strive to **maintain at least the existing ones.** Many a rarity is pursued pretty much without any understanding, and, like it has been prohibited in Switzerland to dig up Edelweiss with the roots (the flower may now only be cut with the knife), so some of our plants deserve a similar protection, preferably without any official police regulation; in some areas such protection is practiced successfully.

With humor and not totally in vain, Count Moltke addressed

the visitors of the landscape park of Kreisau by an inscribed tablet, whose exhortation has stayed in my memory:

"For every foot is every trail,
For every tired person every bench,
For every eye is every flower,
In this nice property;
For heart and mind I dedicate everything to you,
But nothing in here is for the fingers."

Similar exhortations, addressed to the public, also have found observance here.

Chapter Nine

Embellishment of the Forests by Stone Blocks

Chapter Three, Part I, B, discussed the value of stones in the landscape.

It has already been said in the chapter on road layout that particularly beautiful rocks have to be made accessible [so people can enjoy them.] It is rather difficult and expensive, to artificially overcome a lack of beautiful stones, where it exists; but this does not prevent me from giving space to the principles below, for I have often noticed that people experiment with stones in the forest, despite the efforts and costs, and make mistakes in doing so.

The naive beginner, however, has no idea of the difficulty. Not only the average gardener, but also a countryman every now and then does not think of the task as being too difficult. For example, the East Friesian farmer likes to throw up a litte mound in his garden in order to collect on this excellent place erratics [drift boulders]. These are then preferably set up vertically against each other, so that the visitor will see as much as possible from the material brought here.

"A good person"—without doubt the East Friesian farmers

belong to the good people—"is, within his dark [subconscious] urge" at least partly conscious about the right way, inasmuch as **ground modeling** is a prerequisite for effective establishment of stones.

Garden artists, too, can make mistakes. In 1880, R. Geschwind published 346 pages on *The Rocks in Gardens and Parks,* in Stuttgart. His merit lies therein, that he emphasizes taking nature as a teacher: "One should avoid in this" (so he writes in the section dedicated to the establishment of **rocky hills**) "to hesitantly create something unnatural, but one should study the composition of such rocky hills in natural surroundings." But the proof is lacking that he himself has learned from nature in a way others could understand it too. At the same place, he says, for example: "As the **excess** has to be avoided, the **deficient** does damage, too; also one should avoid the **vertical** setting **of all stones,** and one should mix **flat, horizontal** stones together with pointed ones, here sharp edges, there clear level areas; there are cases where even the projecting of larger, column-like formations is quite appropriate."

This will not be of much help for the beginner, even the part that is right, but partly the advice is quite misleading. Deficient will never be able to spoil anything; "setting up vertical" is allowed with stones only as an exception, probably if at all, then, when it can be made believable that the stone had been lifted by a windthrown tree, and had been torn out of its original natural position.

There are even actually incorrect statements, like for example: "The largest stones, of course, form the support of the group." In nature one can find the opposite much more often. For me, living in the diluvium among erratics, the group of erratic blocks stacked into a cairn by Maechtig, the brilliant builder of the Kreuzberg Rocks in the Humboldthain, was particularly interesting and instructive. The artificial arrangement of these rocks can serve as a proof that the largest stones do not, "of course," belong to the bottom.

When the terrain has been chosen correctly for the establishment of the stones or has been modified adequately by the art, then a distinction must be made between **"native grown" stones and**

transported stones. If one has only sharp-edged stones at one's disposal, so only the manifestations of the former ones have to be decisive, as it was explained on page 59; if there are stones with rounded edges only, then the other laws for arrangement have to serve as a guide-line. Whoever has both types at his disposal has to plan the distribution in a way that the sharp-edged ones will be established at the slope and at the bank-edge of watercourses. In the area of the natural or artificial water course, those have to be laid down which more or less clearly show off the traces of transport in water. Sharp-edged stones may be mixed among them, but only individually, as if they had just recently "broken down," as this happens in nature, too.

The correctly chosen stones have to be laid down in the alluvial area in a way that the people think the "currents are responsible for their position; for only in that way we follow nature, connect the stones into a unitary group and give gracefulness to the group.

Stones being graceful ?!—In a certain sense, they really can be; for, as Schiller says rightly, "gracefulness can only be in the movement, but"—so he continues—"this does not hinder, because solid and resting features can also display gracefulness. These solid features were nothing else than movements originally, which finally, with many repetitions, became a habit and left remaining traces."

The fullness of the investigations above frees me from further discussion; for I would only repeat myself if I developed again here the rules of layering, in order to remark with each one: One should also take nature as an model in this.

A part of Moltke's letters first pointed out to me that stones can be used most efficiently in and around the water. Count Moltke writes, in his *Letters From Russia,* on a man-made creek in Peterhof: "What I liked the most in this park, and what surprised me the most at the same time, was a creek, a real German creek with crystal-clear water, which roars along on large blocks of granite. I would not have expected a gradient like this in the level Russia from the Waldai down to sea level.

It has always been incomprehensible to me how the garden

artists of the plain like to lay out waterfalls, instead of using a steep gradient to create a murmuring and splashing creek, at least for a short distance. There, such an artificially tortured watercourse jumps over a board into a six-foot deep abyss, in order to then creep on ashamed, not knowing anymore where to go, if it is not supposed to be up the slope. All we need to know is that the cataract is only let go when the spectator is standing there with raised eyebrows, waiting to be amazed."

Whoever is going to regulate the discharge out of a pond according to the hints on Page 208 in this book, will be able to consider these ideas.

I have been reproached that all this does not belong to *Forest Aesthetics,* but in a textbook of landscape gardening; but I can refer to Burckhardt who assigns to the forester the task "to regulate the spring and the waterfall"; also I can say in my defense that very often foresters actually do not resist the temptation to perform such constructions. However, they are mostly laid out incorrectly, and therefore it has to be taught how to do it better.

The principles I thought to have found still need very much to be checked, and when they are right, the application still remains a very difficult task of artistic tact; furthermore, the moving of large blocks is difficult and expensive. Therefore it is even more important **that one maintains those attractive stones which nature has already appropriately placed,** and has advantageously presented them for view.

If one offers stones for sale, then one should exclude all the exceptionally attractive blocks from the production. Because the workers are very much inclined to provide themselves with a quick progress by dynamiting the largest loose blocks, the stones to be saved have to be clearly marked and numbered, if possible, for as long as the clearings in the district are supposed to last.

When stones of considerable size are found deeply sunk into the soil so that they jut out only a little bit, then they may be lifted upwards. But sometimes it is easier to give a better effect to them by removing the soil from around them.

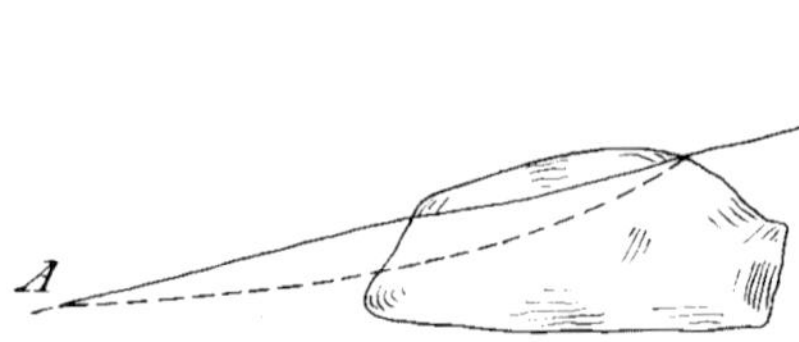

Line A-B is the natural surface. The dashed line represents the newly graded surface.

Fig. 56

In the Schorf Heath, the Hubertus wayside shrine, after which the hunting castle was named, has been erected on an erratic boulder, deeply sunk into the ground, according to a drawing of King Friedrich Wilhelm IV. This simple monument would gain a lot if the stone would be exposed a little more by a little modeling of the ground, according to the pattern in Figure 56.

If stones need to be brought in, then one should choose, if possible, some of picturesque shape and beautiful color, in order to make the effect equal to the effort put into it. If the weight of the chosen blocks is too great, so that the transport as a complete piece is not possible, then they need to be split up and moved as separate pieces. I would excuse it as a permissible deception, if somebody made two or more stones out of one large boulder, in order to use each fraction itself in a way that it will represent a complete erratic boulder, as shown in Figure 57.

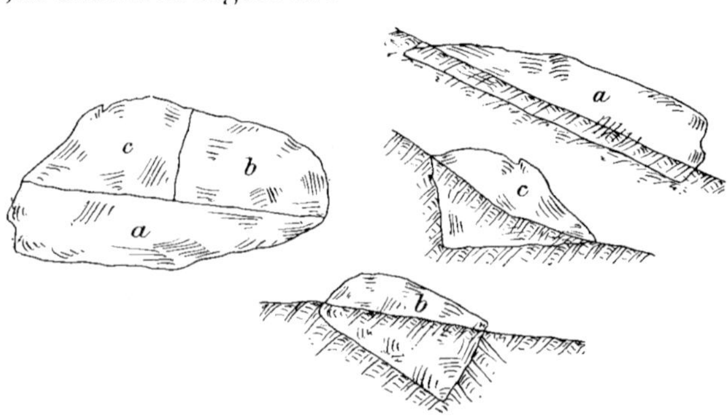

Fig. 57

I think it is less advisable to **put the block together again** into its former shape at its new place by using cement or mortar (which has to be colored as similar as possible to the stone, as Geschwind suggests). If one wishes to raise the mass effect, whose value I realize, then one should restore the former shape without using binders, even if only approximately, so that one may believe that in the course of millennia the stone has been broken by frost.

This appearance, however, can only be created if the stone has been split up according to its **natural veining,** which requires a lot of expertise. This is easier done with wedges than with explosives, whose effect cannot be predicted.

The clefts which remain after joining it together may be filled up with good soil, in order to plant ferns, mountain-ashes, birches, etc. into it. By this, it will be made even more credible that nature itself has widened small existing cracks into extended rifts by using plant roots as a driving force.

In order to make the plants thrive, it is important to consider the water supply. The seemingly natural clefts need to be arranged in a way that water will be collected. So in Figure 58, a schematic sketch, A would be arranged in the right way, B in the wrong way.

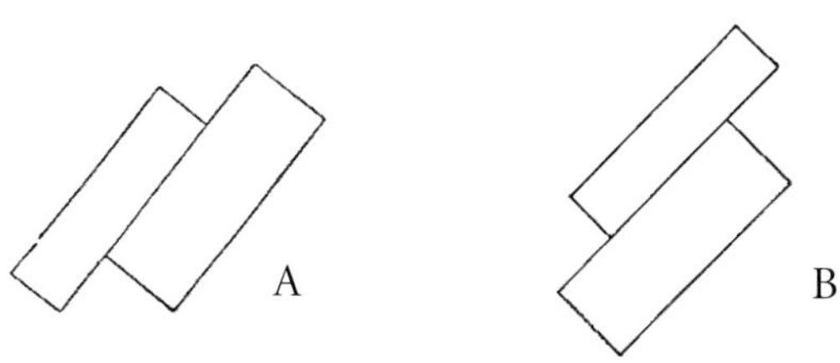

Fig. 58

I owe this hint to Wocke, whose excellent instruction for the arrangement of stone groups should be consulted by everybody who plans undertakings like that on a large scale.

Often **so beautiful or so interesting surfaces of the break** will be revealed to the eye of the surprised viewer that one will not want to hide these. Who would want to bury a gleaming Schrift granite, a tourmaline granite, a porphyry with large crystals, or a rare greenstone, according to Figure 57, in such a way that only the inconspicuous weathered outer surface will remain visible! I could not easily make that decision. Such rare, splendid specimen, do not belong in a stone group. They can be used more effectively in **rough construction of field stones** at a conspicuous place. With this suggestion, I know of a number of lookout towers, war memorials and other monuments built partly or completely out of erratic boulders. But also very modest buildings (like a foundation wall) maintaining the slope at a road cut, or the road sign at an intersection may gain very much if they are built of especially colorful stones.

Chapter Ten

Monuments, Ruins, Entrenchments

The saying, *saxa loquuntur* [the stones talk], is not only valid for the naturally existing stones, but even more for the remnants of human architecture.

Venerable stone witnesses of past centuries have maintained themselves in quite a great number, particularly in the German forests; but well-intentioned eagerness even more than careless indifference has contributed to decrease the charm of these treasures.

This situation has recently improved. The understanding of how to take care of such things has been supported by rich literature. A science of its own has been developed from it.

Goethe details how a **castle ruin** has to be handled in a "Novelle" in the *Unterhaltungen deutscher Ausgewanderter* [Conversations of German emigrants]. The same subject and the importance of architecture in the landscape in general is treated accurately and in the most attractive form by Utis in his novella *Der falsche Baurat* [The False Head of the Planning Department], which I have already cited many times.

One will be able to artificially maintain the less important earth-works (circular mounds, old entrenchments) if one considers their lines in the plantation. One should not draw the planting cord across it, but one should adjust the rows to the remnants of the old works, and one should not miss the opportunity to attract the attention of the visitor to the extraordinary things by using different tree species there. I would like to recommend for places like that the venerable yew tree, which is particularly "idyllic."

Megalithic graves and other smaller but interesting remnants of past times should never be made to disappear completely in thickets. But if one does not want to separate their immediate surroundings from the management system, then selection cutting has to be practiced around them, in order to make the location never appear totally clear nor totally opaque.

Chapter Eleven

Vistas

It is everybody's pleasure to sow, to plant, and to groom, which has been discussed in former chapters, but many do not like to approach the cutting. Quite a few forest owners put off the decision all their lives, and quite a few foresters mark the cutting only unwillingly, forced to do so only by the evil budget plan. Only a few people are aware that cutting, if skillfully limited, often can contribute more to the enhancement of an area than can be achieved with planting of trees. **For it is not at all rare that one cannot see the forest for the trees; often there is lack of the opportunity for an overlook, or a vista which can only be created and maintained by cleverly done cuttings.** Education, training and inherited talent are necessary in order to do this not only well in a craftsmanship sense, but really well. When looking at a really poetic vista that had been opened up by Petzold, at that time Director of the Muskau Park, the wife of General R. von Schlegell said to him: "You have to be a poet." His answer was, "I write poems with the axe." By the way, Petzold has had to face quite a lot of hostility due to his composing with the axe, and the same happens to every landscape gardener when he is forced to remove a tree.

One might remember the temporary storm of indignation that was caused by the cuttings done in the Berlin Zoo. In this situation, we foresters are better off; people are more willing to pardon us for the same thing. But we must never forget to use this advantage knowingly for our purpose; then the **creations of the art of forestry can really often put the park into the shade. Mostly they have the advantage of spatial size over the park, and they always have the advantage of the pleasant idea-connection of utility.** In detail now, I give you the words of Prince Pueckler. The following sentences, taken out of his *Hints for Landscape Gardening,* have, however, been written while thinking of garden and park, but they may claim nearly unchanged validity for the conditions of the art of forestry; I quote:

"Everything in the more distant landscape which may offer anything interesting will be, so-to-say, drawn into our property; all the more distant rays will be focused to this point, and therefore an imaginary boundary will be created which may exceed the real one infinitely, if it is used skillfully. These distant points have to be handled in a way that one will never sensually realize the boundary between them and one's own position, even if one can imagine that they are too far away to be still within the real range of the grounds. As few as possible of the distant points should be visible from other viewpoints. For example, mountains always should be made partially visible, and only once should they be seen in their whole extension; a city should be subdivided in the same way, and one should not let the same object reappear several times. The effective hiding and leaving-to-be-guessed is more difficult than the open showing. When the viewers think of a view as being amazingly beautiful and then, later, after a longer contemplation, utter: What a pity, that the large tree over there is still standing in front of it, how much more magnificently everything would unfold if this one were gone, too; then one has hit it just right, as a rule, and the good people would be very surprised if one did them this favor and took away the condemned tree, too, and then, all of a sudden, they would not see any picture any more, for a garden in large scale is only just a gallery of pictures, and pictures demand their frames. . . .

Buildings never are supposed to be shown totally freestanding; otherwise they will appear like spots and are standing there as strangers, not grown together with nature. Half-hidden things, anyway, are advantageous to any beauty, and in this area of imagination there is always something left to be guessed. Often the eye rests with more pleasure on a mere chimney in the distance, which whirls its gray smoke clouds from the unending forest area up into the blue ether, than on a bare palace, which is accessible from all the sides, does not offer even one single living interruption to the eye, and to which nature has not yet snuggled up lovingly and homely at any place."

Out of my own experience, I add some remarks about middleground and foreground to this quotation (it teaches in detail the use of the **background**):

It is most important for the **middleground** to have an open and **stable grouping** of the masses. This will need to be **considered** especially **in the seed-tree system. However, one will have doubts sometimes for economic reasons to unite into groups the trees which are supposed to be left as standards, but these mostly will be allowed to be arranged in irregular rows in such a way that, when seen from the most important viewing points, they appear close together.**

When they lack adequate arrangement into groups, even the most beautiful trunks do not really show to their advantage, from my local experience. Quite a number of the mightiest trunks had been saved in the "Kaelberwinkel" at the forest's edge in Postel as a remnant of a valuable stand of old oaks. But they formed only a row full of gaps, and neither they nor their background showed to real advantage. Then, I decided to eliminate the most isolated trees and now the remaining ones order themselves quite naturally into three strong groups to the viewer's eye, and stand out in a magnificent way as dark masses against the pale background of the vigorous young trees.

I ask the kind reader to remember the proof offered in Part One, that the human does not only see with the physical eye, but also with the mental eye.

This can be observed especially at vistas. Whoever enters the flat roof of the Johanna's Height [lookout tower], usually is first enchanted by the forest scenery, lying at his feet, and enjoys a purely aesthetic pleasure. But soon the stranger will demand all sorts of information—he wants some orientation. "Look," he is told, "there lies the Trachenberg and Winzig, here is the Zobten, over there the Schneekoppe"—and if the guest has discovered these and other noticeable points, so this will increase his pleasure, although he has not really seen these towns except for their church steeples, and instead of Zobten Mountain only a bluish mist. The Schneekoppe usually can be noticed only with particularly clear air. Mostly, the stranger has to believe his guide that it does appear every now and then, and he has to complete by imagination what he cannot see physically. But the **orientation,** good for the enjoyment of a landscape, is nowhere so difficultly obtained as in the monotonously rolling landscape, where the road, be it a man-made street or be it railroad, leads up and down in endless curves. This is nearly like on the sea, when the ship aims for a difficult shipping canal between flat islands. Then passengers as well as crew look attentively for the lighthouses and buoys which mark the islands and promontories and they who have recognized only the buoy are as delighted as if they had seen the whole island. In a similar way in the hilly region, the hiker looks for single church steeples, groups of trees or large trees known to him, in order to orient himself in the area.

The forester can best approach this need by groups of standards.

But it is nearly always a mistake to save individual trees on the tops of hills as a symbol of a whole area. Being unprotected against the rigors of the weather, they will soon give an impression of being lonely, causing pity, and sadly, most will die after a few years.

This has already been recognized by Isaiah. He describes the sad fate his contemporaries are hurrying towards with the words, "One thousand shall flee at the rebuke of one; at the rebuke of five all shall flee: till ye be left as a beacon upon the top of a mountain, and as an ensign on a hill." [Isaiah 30:17]

So it is particularly advisable on mountain tops not to retain too few standards. I remember the Elfbuchenberg ["Mountain of Eleven Beeches"] in the Royal Forest District of Ehlen, in Kirchdittwald, near Kassel, as being a good example. As the name says, eleven saved beeches adorned the mountain and by this, the whole area.

The eagerness of many Associations for Embellishment goes too far nowadays; they seem to have set up as their task to mark all the remarkable tops of mountains with high **lookout towers.** A lookout tower immediately adjacent to a town, like the one near Eberswalde, will not disturb anything. But in the heart of the forest, one should limit the eagerness of the societies and the architects. The architecture must not want to push itself beyond the practical need; therefore the towers must not stand out considerably above the surrounding forest. So the Elisabeth Tower near Eutin in the beginning was built only high enough to tower above the surrounding crowns of beeches, and only later, when the beeches were growing up, was it extended upwards.

Tall lookout towers must not be released by later clearcutting. I use this rule here in the way that I separated 5 ha of selection forest near Johanna's Height, although this building, laid out on a wide surface would tolerate a release better, and it represents more than only a lookout tower. So, Pfeil's beautiful poem, which I put in here, does not refer to a modern lookout tower, but to an inhabitable high hunting lodge:

Deep in the silence of the beech forest,
There is a small narrow house,
Which looks out, sheltered by high oaks,
Far into the blue distance.

Boldly the structure rises out of the trees,
At its foot the green forests are lying,
The water can be heard foaming below,
The mountains can be seen glowing in the evening.

This house guarantees in its narrow rooms
The most beautiful clear hunter pleasure,
And if I can dream myself there,
Then longing often swells in my breast.

In later years, Pfeil built next to his hunting lodge, which he had erected instead of a tower, a tree-stand in an oak by fixing securely a sturdy balcony up in the crown and making it accessible by a stable ladder.

In someone else's districts, I have very often been asked about plantation areas, whether existing **advanced regeneration groups** are to be maintained or to be removed. Such stands of advanced regeneration are often of high aesthetic value on large clearcut areas, because they serve as **middleground,** and their removal easily destroys the most beautiful scenes. Therefore, for advanced regeneration groups, the same is valid as for the standards: They have to be distributed in a picturesque grouping in order to be beautiful.

The photograph put in here (Fig. 59) unfortunately doesn't show reality, because the camera has reproduced very nicely the middleground, but not the mature wood stand in the background, the view of which is framed by the advanced regeneration groups.

We have even more influence on the **foreground** than on the middleground; unfortunately, this power is abused sometimes. At least I have seen several times where vistas were kept free by **cutting off the crowns of trees.** I have previously remarked that is an impertinent method when discussing the selection forest, and here I add that a conservative cutting, not limited by schematism, can replace an overgrown vista. This **gradual change of scenes** is another advantage that the forest has over the park.

Fig. 59

Where the terrain drops off rather steeply, it is advisable to keep **permanently** free a certain vista **also in the interior of the forest,** without being forced to use this frowned-upon method (the beheading of the trees), if one can decide to tolerate another small sacrifice: Namely, where the terrain is rocky, I recommend to expose the next 15 m of rocks downslope by completely removing the soil, so that it will look as if nothing besides ferns and mountainashes has ever been able to find a modest little foothold there. Further down the slope, it then has to be cut after the selection method if necessary; that means, trees whose crowns cover the background in a **disturbing** way have to be cut out early. But where the soil does **not** possess a shallow stony foundation, the gap in the stand has to be handled in a way that it appears as if it had been created by snow breakage or other circumstances out of the control of the manager. Hazels, alder, and other low shrubs are to be established soon, if they do not volunteer satisfactorily. This method will look the most natural where the view **below the crowns of old trunks** is to be kept open, for their shade will prevent rampant young growth so that the pruning shears will not need to be used. Such a framing of scenes from above is also very well liked by landscape painters. How often one can see Rome represented with such a foreground (seen from Monte Pincio). These paintings of Rome all have one thing in common—one cannot really see Rome in them; the holy city is only suggested by the outlines of some main buildings, but the foreground is decorated by a mighty stone bowl beside large trees with wide crowns. Now, we cannot match this in the forest, but the natural treasures of the district, being stone blocks, often may serve for the same purpose. Especially the **drift boulders or erratics,** of the plain and the hilly region are a very useful adornment of the foreground, which can even be increased by planting beautiful shrubs (Rose, European Cranberrybush [*Viburnum*], etc.).

Long narrow valleys and water surfaces present a particular difficulty when they have nearly straight boundaries, for they open up a wide view only from their ends. At three places, I have met exemplary installations that helped overcome this problem. Near Lauterberg in the Harz, I found, beside the road which ran along the bottom of a valley, an artificial mound erected like a bastion, in order to create the level area required for a charcoal pile, which could not be obtained otherwise. From this place [the mound], it was possible to look very nicely up and down the valley. Something like that can be created occasionally in road construction work without considerable additional costs, and the small effort will be rewarded abundantly by providing for quite a few conveniences (rest area for horse and cart, area for turning-around, place for timber storage). Only for the reason of beauty, Wilbrand has created something similar near Darmstadt. Beside a creek, the Darm, a promenade road runs in the forest beside a meadow's edge, which is widened at the side of the valley by piling up some so-called pulpits at several points. These pulpits are secured by dry walls with rather steep walls at the side of the creek. On top, they are bordered by a breastwork of peeled oak timbers. The view from these pulpits pleases by contrast the more, as this road is led otherwise within the stand in a way that the views across the meadow valley can only be enjoyed freely from these points, according to a rule suggested by Wilbrand: "Foot paths for promenading are to be laid out near towns within but close to the forest's edge, in order that between the path and the forest's edge there is a narrow strip of woods several meters wide. If the trees in the strip retain their lower branches all the way to the ground, then the wall of leaves will shine to the one looking into the direction of the sun in a bright, shiny golden finish."

At the water, landing stages and breakwaters are the ones which, besides practical tasks, also serve the purposes of beauty in opening up vistas. At the same time, they enhance the water by interrupting the often monotonous lines. The small artificial spit of land at the hunting castle Grunewald serves both purposes very beautifully.

I would like to finish this chapter by a general contemplation taken from Gilpin:

The appreciation of a vista is not only dependent on the landscape, but also on the present state of the viewer. One is not always in the mood to have magnificent views work on oneself. The tired body, even more the tired mind, will prefer a certain limitation.

For a careful treatment of the enhanced forest, the rule can be deduced from this, that very different functions have to be provided for the different needs. If it is about magnificent scenes, then one will do well to prepare for the great impression by smaller views, which the hiker can take on the road. If the landscape offers minor attractions, then one will need to present the most beautiful, the most interesting one has in a way as surprising as possible. Prince

Pueckler used this strategy in the Branitz Park. When I visited this most ingenious creation of the art of gardening many years ago, I was surprised in the beginning by the monotony of the "perimeter road," whose edges were bordered by such dense plantings that not even the slightest view could be seen. Then a stone bench and a stone table became visible to attract my attention, and when arriving there, I found a vista kept free to the top of a large pyramid, built up at the artificially dug-out branch of the Spree, which the creator of the park had piled up at his monument.

One will not often be able to imitate this in the forest. Be it the view onto a new water course, onto a beautiful tree, onto a forester's house, etc.: if it is to be appreciated adequately, then not only the scene but also **the viewer will need to be prepared accordingly.**

Not only in this respect, but quite in general the exhortation is valid: **The forester is not only supposed to groom the forest beauty, he is also supposed to condition the forest visitors in a way that they are able to understand and to appreciate the offered magnificence of the forest!**

Appendix

Explanation of the Pictures

Photographs [Plates]

Plate I. P. 55. Torrential stream, carrying detritus, boulders, silt.—After a photograph given to me by the art trader Ed. Schulte, Berlin, of a painting by Carl Hasch (died in Vienna 1897), representing a scene along the way up to the Stuiben Falls near Umhausen in the Oetzal [Austrian Alps].—Refer to Page 63.

Plate II. P. 65. Susanna Oak.—After a photograph by Teacher Wassdorff in Postel. The diameter at breast height is 185 and 202 cm, the circumference is 6.73 m.—The Susanna-Oak belongs to the saved standards at the edge of the Kaelberwinkel in Postel.—Refer to Page 209.

Plate III. P. 129. Nesigoder Zoo, island with alders.—After a photograph by Wassdorff. In the so-called Luge, there are numerous such islands, which may even be called floating, as they start to sway with heavy wind, being only grounded by alder roots.—See also Page 130.

Plate IV. P. 135.**Postel, Danckelmann-line.**—After a photograph by Wassdorff. The perspective effect has been increased by narrowing the firebreak in the background.—Refer to Page 136.

Plate V. P. 157.**Postel, standards in naturally regenerated pine thicket.**—After a photograph by art teacher Peltz in Breslau. The picture shows regularly [evenly] branched pine standards, united into groups, in Compartment 52.—Refer to Pages 156 and 180.

Plate VI. P. 163. Postel, old pollarded willow tree. -After a photograph by Wassdorff.—Refer to Page 162.

Plate VII. P. 182. Postel, advanced regeneration of oaks, in groups, with pine standards [shelter trees].—After a photograph by Wassdorff. The picture shows a regeneration group west of the "long line" in Compartment 56 a. By the removal of encroaching trees over a period of 20 years, the middle-trunk has been released and the growth of young oaks by seed has been favored, to which spruces have been added as a nurse tree by planting.—Refer to Pages 180 and 186.

Plate VIII. P. 190. **Postel, Postel thinning in a beech stand.** After a photograph by Wassdorff. The small trunks visible in the picture, only a few centimeters strong and forming the understory, are about even-aged with the dominant stand. (One of the oldest experimental areas between Johanna's Height and Dachsberg.)—Refer to Page 189.

Plate IX. P. 197. **Oaks on the Militsch Road.**—Photograph by Wassdorff. The oaks planted as young seedling trees in 1877 and have been groomed since then (for 23 years) by pruning.—Refer to Page 195.

Plate X. P. 204. **Prince Forest near Ohlau, the Weinert-Meadows**—After a photograph by J. Volpert in Ohlau. (Refer to "Bunte Bilder aus dem Schlesierlande" ["Colorful Pictures out of Silesian Land"], published by the Silesian Pestalozzi-Society, Breslau, 1898.) The picture shows beautiful views and variously formed forest edges.—Refer to Pages 205.

Plate XI. P. 206. **Nesigoder Zoo, bank of the Luge.**—After photograph by Wassdorff. The picture shows the Luge above the island-terrain, shown in Plate III. Notice the development of the alders at the bank, which determines the course of the creek.—Refer to Page 207.

Plate XII. P. 226. **Kratzkau, view into the open landscape.** After a photograph by Forest Trainee von Salisch. The picture shows the castle meadow, decorated by standards and group-plantings according to the plan of Lenne. In the middle-ground two oaks are standing, which have been pruned according to advice from Count Moltke; before that, following the hedges of hawthorne which are kept low and border the public road, there is a group formed by *Cornus alba* in order to decorate the scene with color in winter.—Refer to Pages 231 and 247. (Hedge in place of an allee!)

Plate XIII. P. 230. **Postel, wild pear tree at the village green.**—After a photograph by Wassdorff. The picture shows the tree decorated by hoarfrost. On the side of the tree, turned away from the viewer, there are benches, installed for the school children below the crown of the tree.—Refer to Pages 82, 228.

Plate XIV. P. 244. **Berlin Zoo, allee at the Flora-Place.**—After a photograph by Court Photographer Braatz, Berlin. The picture shows a clipped yew bush in the intersection of allees.—Refer to Page 241.

Plate XV. P. 270. **Postel, Emily-Beech.**—After photograph by Wassdorff. The picture illustrates that wide-crowned trees lose their peculiar character in the high forest, because the lower branches are shaded out by the fast-growing younger trees. Refer to Pages 267.

Plate XVI. P. 280. **Landscape in the Rhine Valley, view from the Island Moenchsau onto Reichhartshausen**—After a photograph given to me by DR. jur. Wilhelmj, of the Court Photograph C. Colberg in Bad Oeynhausen. The picture is meant to prove that pyramid trees have to stand in groups and they look especially good at the water and in front of ranges of hills.—Refer to Page 279.

Figures and Pictures in the Text

Fig. **1,** P. 26. Geometric figure, the **golden section.**

Figs. **2** through 7, P. 57 through 61. **Layering of stones.**—Drawings by the author.—Refer to Page 284.

Figs. **8** through **16,** P. 71. **Oak leaves,** drawn from nature by Johanna von Salisch.—Refer to Pages 70 and 72.

Figs. **17** and **18,** P. 72 and 73. **Oak leaves,** picked and drawn by King Wilhelm I. From copies given to me by Major v. Meyerlink.

Fig. **19,** P. 93. Pine with hanging branches in Compartment 167, Royal Forest Administration Katholisch-Hammer.—From photograph by Wassdorff.

Figs. **20** and **21,** P. 102 and 103. *Betula pendula (verrucosa)* and *B. pubescens (odorata).* Drawn according to nature by Johanna von Salisch.

Figs. **22** through **27,** and Fig. **30,** P. 138–140, 143, and 145. Figures concerning the **road network,** drawn by the author.

Fig. **28,** P. 144. **Vista** opened up by clearcutting in Compartments

63, 89 and 90, Royal Forest Administration Katholisch-Hammer.—From photograph by Peltz.

Fig. **29,** P. 145. Schematic presentation of the development of views in the high forest.

Fig. **30,** (see Figs. 22–27).

Fig. **31,** P. 164. **Lopping woods** in Zwornogoschuetz, District of Militsch. After drawing of the landscape painter A. Bethge, Berlin.

Fig. **32,** P. 167. Area in the Langengrund, **Riesengebirge [Sudeten Mountains].**—After a photograph by A. Fabian & Co., Breslau, given to me by the company of C. Schroeder in Breslau. The picture shows the beautiful contrast of pointed-crown trees in front of a long stretch of mountain range.

Figs. **33** through **35,** P. 184. Schematic drawings of **cultivation areas [plantations].** (Refer to Page 184.)

Fig. **36,** P. 187. **Oak stumps** from the Spitzberg [Spitz Mountain], Royal Forest Administration Katholisch-Hammer, after photograph by Wassdorff. Last remnant of the "Paradise" in those compartments. (Refer to P. 154 and 187.)

Figs. **37** and **38.** P. 208. Schematic drawing of the **pond stoppers** used in Postel.

Fig. **39,** P. 209. **Edge of a forest meadow [pasture],** formed like a selection cutting, in Katholisch-Hammer, Compartment 90. After photograph by Wassdorff.

Fig. **40,** P. 210. The **beechbank beeches** in the Postel Muellerheege. After photograph by Wassdorff.

Fig. **41,** P. 212. Schematic drawing of a **hedge planting** after drawing by District Forester Conrad in Klein-Commerowe.

Fig. **42,** P. 222. The **Johanna's Height** in Postel. After photograph by Peltz.—Refer to P. 222 and 294.

Fig. **43,** P. 224. **Susanna-Pines** in Postel, Compartment 48 b. -After photograph by Wassdorff. The pine stand has been severely thinned repeatedly (selection thinning in the way described on P. 191–192). The development of oaks on the western side has been encouraged by

removal in strips at the edge. The pines, which have been prepared very gradually for free position by cutting their crowns free, have not suffered any noticeable storm damage.

Fig. **44,** P. 233. Schematic drawing of the method of cutting on long hills.

Figs. **45** and **46.** P. 239 and 240. Schematic drawings of the **embellishment of rectangularly intersecting firebreaks.**

Fig. **47,** P. 242. Oak, planted at the boundary of Postel and Protsch, according to the pattern of Fig. 45, C.—After drawing by Ms. Edith Schiemann, Berlin.

Figs. **48** through **52,** P. 247 through 252. Schematic pattern for road borders and allees.

Figs. **53** and **54,** P. 255. Fruit trees with regular built-up crowns according to Gaucher, "The Graftings," and "Practical Fruit Cultivation," 1885 and 1891, [respectively,] the former published by Hoffmann, Stuttgart, the latter by Parey, Berlin.

Fig. **55,** P. 265. Trunk, laid up with bricks within, after Wilbrand.

Figs. **56** through **58,** P. 288–289. Schematic drawing of naturally layered stones (drawing by the author).

Fig. **59,** P. 297. Advanced regeneration groups in the Royal Forest Administration Katholisch-Hammer, Compartment 89.—After a photograph by Wassdorff.

Used Sources

Before the listing of the printed sources used by me, I have to state my gratitude for oral or handwritten information given to me for supporting my efforts. D. Kleinert, professor and Royal Oberkonsistorialrat in Berlin, has supported me in the general part already when I was writing the first edition; von Bornstedt, now Head Forester in Berlin, has supported me for the applied part with literature sources and good advice. For the second edition, Rodig, Royal Head Forester, previously in Katholisch-Hammer, now in Jellowa, has provided me with his valuable assistance.

Remarks to PART I, Section A.

Chapter 1.

To 1.

To Pages **1 and 2.** Krause's "Landverschoenerkunst" [The Science of Landscape Art] has been edited by Dr. Hohlfeld and Dr. Wunsche, by Otto Schulze in Leipzig, now with the publisher Emil Felber, Berlin. Besides this one, I have used: Krause, "Vorlesungen ueber Aesthetik" [Lectures on Aesthetics] and "System der Aesthetik" [System of Aesthetics], Leipzig 1882.

To Page **2.** Dr. K. E. Schneider, "Die schoene Gartenkunst" [The Beautiful Art of Landscape Gardening], Stuttgart 1882, and: "Ein Gaertner als Aesthetiker" [A Gardener as Aesthetician], Dresden 1884.

To Page **3.** Vischer, "Aesthetik Teil II, Die Lehre vom Naturschoenen" [Aesthetics Part II, The Teaching of Beauty in Nature], Leipzig 1847.

To 2.

To Page **3 and 4.** Krause "Vorlesungen" [Lectures], page 187.

To Pages **5.** G. Heyer, "Supplement zur Allgemeinen Forst und Jagd zeitung X" [Supplement to the General Forest and Hunting Journal], page 26.

To Pages **5 and 6.** Guse, "Waelder oder Gelder" [Forests or Money]. Forstliche Blaetter [Forest Information] 1887, page 200.

To Page **6.** Guse, "Die Anwendung der Reinertragstheorie auf die Staatswaldungen" [The application of the theory of the highest net profit on the state forests]. Muendener forstliche Hefte [Muenden Forest Journals] 1899.

To Page **7.** von Salisch, Zeitung fuer Forst and Jagd [Journal for Forestry and Hunting] 1892, "Die Beziehungen zwischen dem Schoenen und dem Nuetzlichen im Forstwesen" [The relations between the Beautiful and the Useful in Forestry].

To Page **8.** Pfeil, Kritische Blaetter [Critical Pages], 37 II, page 197.

To Page **8 and 9.** Forester Roedler, Deutsche Forstzeitung [German Forest Journal] 1901, page 376.

To Page **9.** Neumann, "Was ist die Kiefer?" [What is the Pine?], Zeitung fuer Forst und Jagd 1890.

To Page **10.** Count von Moltke, *Briefe* [Letters], Berlin 1891, page 23.

To Page **11.** Schulze, *Jahrbuch des schlesischen Forst Vereins* [Yearbook of the Silesian Forest Society] 1879, page 89.

To 3.

To Page **13.** Masius, "Naturstudien" [Nature studies], and: J. Fischbach and H. Masius, "Deutscher Wald und Hain in Wort und Bild" [German Forest and Grove in Word and Pictures], Munchen, Friedrich Bruckmann (a magnificent book, without the year of printing!).

To Page **13.** Rossmaessler, "Der Wald" [The Forest]. I quote from the First Edition. The Third Edition has been published by Willkomm.

To Page **13.** Gayer in F. Zbl. 1897, pages 314 and 319.

To Page **14.** Willbrand, "Forstaesthetik in Wissenschaft und Wirtschaft" [Forest Aesthetics in Science and Management], *Allgemeine Forst und Jagd Zeitung* 1893.

To Page **15.** Ritter von Guttenberg, "Die Pflege des Schoenen in der Land-und Forstwirtschaft" [The Care for the Beautiful in Agriculture and Forestry], *Oesterreichische Forst-Zeitung* [Austrian Forest Journal] 1889.

Head Forester Leuer, "Waldaesthetik und Fremden-verkehr" [Forest Aesthetics and Tourism], *Darmstaedter taeglicher Anzeiger* [Darmstadt Daily Advertiser] 1893, June 18.

Dr. von Fischbach, "Einige Vorschlaege zur Waldver-schoenerung" [Some Suggestions for Forest Enhancement], *Centralblatt fuer das gesamte Forstwesen* [Central Paper for the Entire Forestry], Vienna 1893.

Kraft, "Zur Aesthetik der Park- und Waldwirtschaft" [On the Aesthetics of Park and Forest Management], *Zeitschrift fuer Forst und Jagd* [Journal for Forest and Hunting] 1895.

Upper Chamber, *Zeitschrift fuer Forst und Jagd,* 1895, pages 326 and 332.

Saechsischer Forstverein [Saxonian Forest Society], 41. Versammlung [41st meeting]. *Bericht* [report] pages 56 until 71.

Forestry Advisor L. Hampel, "Die Vereinigung des Wirtschaftlichen mit dem Schoenen im Walde" [*The Unification of the Economical with the Beautiful in the Forest*], Oesterreichische Forst- und Jagd-Zeitung [Austrian Forest and Hunting Journal], 1889, May 20.

Forest Supervisor Heinrich Fuerst, "Wie vermoegen wir die Naturschoenheiten unsrer Kurorte und Sommerfrischen zu foerdern?" [How can we support the nature beauties of our spas and summer resorts?], Oesterreichische Forst- und Jagd-Zeitung 1898, August 5.

The forest aesthetic essays published before 1885 have been used for the first edition of my *Forest Aesthetics* and they are listed on page 226 and following [in the first edition; see Appendix E for translation].

Chapter Two.

To 1.

To Page **15.** Krause, "System," paragraph 7.

To Page **16.** Darwin, *Band II* [Volume II], page 375.

To Page **16.** Rossmaessler, page 23 of the First Edition.

To Page **18.** Th. Hartig, *Allgemeine Forst und Jagd Zeitung,* 1879, Page 268.

To Page **19.** Oerstedt, "Der Geist in der Natur" [The mind {spirit} in Nature], Leipzig 1854.

Jungmann, "Aesthetik" [Aesthetics], Freiburg im Breisgau 1884, Band II [Volume II], page 58.

To 2.

To Page **20.** Schinkel, compare Utis, "Der falsche Baurat" [The False Head of the Planning Department], page 85, Frankfurt am Main 1877.

To Page **22.** Socrates, the quotation can be found in Jungmann, page 40.

To Page **22.** Shakespeare, *Othello,* I. Aufzug, 3 Szene [Act 1, Scene 3.]

To Page **22.** Wingolf, 3rd song [verse].

To Page **23.** Bruecken, compare with Goethe, "Der Triumph der Empfindsamkeit" [The Triumph of Sensitivity], 4th Act.

To Page **23.** Fechner, "Vorschule der Aesthetik" [Pre-study of Aesthetics], Leipzig 1876.

To Page **25-27.** The theory of the golden section is developed rather in detail in Zeisig, "Aesthetische Forschungen" [Aesthetic Research], Frankfurt am Main 1885.

To Page **28.** Selenka, "Der Schmuck des Menschen" [The decor of Humans], Berlin 1900.

To 3.

To Page **31.** Fechner (at the place already mentioned) has particularly developed the theory of the "aesthetischen Associationsprinzip" [aesthetic principle of association].

To Page **33.** von Riesenthal, Trier, Lintz Bookstore.

TO PART I, Section B.

To Page **34.** Besides the works mentioned at certain places, the following books deserve special attention: Vischer, "Aesthetik" [Aesthetics], Part II, Leipzig 1847; Berthold, "Das Naturschoene" [The Beauty in Nature], Freiburg im Breisgau 1875; Hallier, "Aesthetik der Natur" [Aesthetics of Nature], Stuttgart 1890.

Chapter One.

To Page **34-36.** Hermann, "Die Aesthetik in ihrer Geschichte und als wissenschaftliches System" [Aesthetics in its History and as a Scientific System], Leipzig 1876.

To Page **36.** Jungmann as mentioned above, page 169.

Chapter Two.

To Page **38.** For detailed study of the theory of colors, I recommend Berger, "Katechismus der Farbenlehre" [Catechism of the Theory of Colors], Leipzig 1898. Within that book more literature is mentioned. The work of Petzold, "Farbenlehre der Landschaft" [Theory of Colors in the Landscape], Jena 1853, is remarkable. I received some particular information about the "seeing" from Helmholtz ("Populaere wissenschaftliche Vortraege" [Popular Scientific Talks]), Braunschweig 1876.

To Page **45.** Hermann, as mentioned above, page 201.

Chapter Three.

To Page **54.** Selenka, as mentioned above, page 70.

To Page **54.** Wahnschaffe, "Die Ursachen der Oberflaechenge-staltung des norddeutschen Flachlandes" [The Causes of the Relief Structure of the North German Plain], Stuttgart 1901.

To Page **56.** Moltke, *Briefe aus Russland* [Letters from Russia].

To Page **57.** Wang, "Ueber die Gesetze der Bewegung des Wassers und des Geschiebes" [On the Rules of the Movement of Water and of Detritus and Boulders], Vienna 1899.

The remarkable **erratic boulders** of East Prussia have been described by Dr. Jertsch in the "Merkbuch" [Notebook] for the Province of East Prussia (*Beitraege fuer die Naturkunde Preussens* [Contributions for the Natural Science of Prussia], published by the physikalisch-oekonomischen Gesellschaft [Physical-Economical Society] in Koenigsberg, Issue 8).

Chapter Four.
On the Title Page 63.

Bratranek, "Beitraege zu einer Aesthetik der Pflanzenwelt" [Contributions to an Aesthetic of the Plant World], Leipzig 1853.

Jaeger, "Deutsche Baeume und Waelder" [German Trees and Forests], Leipzig, Karl Scholze.

Schmerz, "Naturgeschichtliche Charakterbilder" [Characteristic Pictures of Natural History], Leipzig 1879.

To 1.

To Page **64. Forms, varieties and subvarieties** are listed rather completely in Lauche, "Deutsche Dendrologie" [German Dendrology], Berlin 1883, and Beissner, "Nadelholzkunde" [Science of Conifers], Berlin 1891.

To Page **64.** Oerstedt, "Geist in der Natur" [Mind {Spirit} in Nature], 1854.

To 2.

To Page **70.** G. L. Hartig, "Lehrbuch fuer Foerster" [Textbook for Foresters], 5th edition, page 65.

To Page **71.** (Figures). von Ettingshausen, "Ueber die Nervation der Blaetter bei der Gattung Quercus" [On the Veining of the Leaves of the Species Quercus], Vienna 1895.

To Page **72.** G. L. Hartig as mentioned above, page 72.

To Page **74.** Petzold, "Die Mutter unserere Pyramideneichen und ihre aelteste Tochter" [The Mother of our Pyramid Oaks and its oldest Daughter], *Wiener Obst- und Gartenbau-Zeitung* [Viennese Journal for Fruit and Garden Cultivation], Vienna, Faesy & Frick 1876.

To Page **78.** Blood beech. "The mother of all blood beeches," which has been known already by Bechstein, was described by Dorl in the *Allgemeine Forst und Jagd Zeitschrift* 1877. At that time, its age was estimated to be about 200 years. It was 27 m high, richly branched, very full-boled.

To Page **80.** Dr. C. Bolle, "Die alte Tegeler Baumschule" [The old Tegel Nursery], *Mitteilungen der deutschen dendrologischen Gesellschaft* [Information of the German Dendrological Society] 1898.—All the words in this section in quotation marks stem from Dr. Bolle, unless marked otherwise.

To 3.

To Page **86-87.** The sentence in quotation marks has been taken from Vischer, "Aesthetik" [Aesthetics], "Die Lehre vom Naturschoenen" [The Theory of the Beauty in Nature].

To Page **89.** The Scheppe-Allee is described more in detail in the First Edition of *Forest Aesthetics,* page 73.

To Page **89.** Lieper Revier [District of Liepe], Grunert, *Forstliche Blaetter* [Forestry Papers] 1878, page 85.

To Page **89.** Kessler, Royal Forest Supervisor in Colpin.

To Page **92.** Karl Koch, "Vorlesungen ueber Dendrologie" [Lectures on Dendrology], Stuttgart 1875, Page 332.

Schmittspahn, Allgemeine Forst und Jagd Zeitung 1885, page 397.

To Page **95.** Dr. C. Schroeder, "Ueber die Vielgestaltigkeit der Fichte" [On the Multiformity of the Spruce], Zurich 1898.

To Page **95.** Haselfichte. Dr. Wurm, "Waldgeheimnisse" [Forest Secrets], Stuttgart 1895.

To Page **96.** Conwentz, "Beobachtungen ueber seltene Waldbaeume in Westpreussen" [Observations on Rare Forest Trees in Western Prussia], (in *Abhandlungen zur Landeskunde* [Discourses on Regional Studies], Danzig 1895) and *Forstbotanisches Merkbuch* [Forest Botanical Notebook], I. Prov. Western Prussia, Berlin 1900.

To Page **95.** J. Fischbach and H. Masius, "Deutscher Wald und Hain in Bild und Wort" [German Forest and Grove in Word and Picture], Muenchen and Berlin by Friedrich Bruckmann (without date of printing).

To 4.

To Page **99.** Lohr, "Die Linde ein deutscher Baum" [The linden, a German Tree], Spandau at Gustav Schob 1889.

To Page **99.** Critic in the *Nationalzeitung* [National Newspaper] 1885, page 396.

To Page **102. Birch-species:** Up to now, I have not been able to find a better study on our birch species than the one written by Th. Hartig. ("Vollstaendige Naturgeschichte der forstlichen Kultur-pflanzen Deutschlands" [Complete Natural History of the Forestry Cultivation Plants of Germany], Berlin 1851.)

To Page **104.** Dr. C. Bolle, "Andeutungen ueber die freiwillige Baum- und Strauchvegetation der Provinz Brandenburg" [Hints on the Voluntary Tree and Shrub Vegetation of the Province of Brandenburg, Berlin 1886, Maerkisches Provinzial-Museum.

To 5.

To Page **106.** Rose. Carus Sterne (Leipzig, Gustav Freytag, 1884 and

1885). I mention summer flowers, fall and winter flowers explicitly at this place, because he treats the wild roses particularly in detail.

To Pages **108.** Pueckler, *Briefwechsel* [Exchange of Letters], Band IV [Volume IV], Berlin 1874.

Chapter Five.

To Page **114.** Finch's Song: On birds singing also see: Altum, "Der Vogel und sein Leben" [The Bird and its Life], Muenster in Westphalia, 1898, and Rausch, "Die gefiederten Saengerfuersten" [The Feathered Princes of the Singers], Magdeburg 1900. (The latter book discusses in particular detail the singing of the native birds).

To Page **116.** Schleiden, "Studien" [Studies], Leipzig 1857, and: "Baum und Wald" [Tree and Forest], Leipzig 1870, page 5.

TO PART II, Section A.

Chapter One.

To Page **119.** Social politicians: E. M. Arndt, "Der Waechter" [The Guardian], Band II [Volume II], page 1815: "Ein Wort ueber die Erhaltung der Forsten and der Bauern im Sinne einer hoeheren, das heisst menschlichen Gesetzgebung" [A Word on the Maintenance of the Forests and the Farmers in the Sense of a Higher, Meaning Human, Legislation]. (Other social politicians who consider forestry points of view are mentioned in the First Edition of the *Forest Aesthetics,* Note 28).

To Page **120-121.** Riehl, "Land und Leute" [Country and People], (Stuttgart). Holtei has expressed the same idea in poetic form and excellently in "Stimmen des Waldes" [Voices of the Forest], Prologue, Breslau 1854.

To Page **121.** Schleiden, "Fuer Baum und Wald" [For Tree and Forest], page 3.

To Page **122.** "Behoerden" [administration], von der Borne, Denkschrift [Memorial Writing], concerning the forest situation in the provinces of West and East Prussia, the decline of the forest in these parts of the country and the means applied and to be applied by the state, in order to stop the evil of advancing deforestation. Danckelmann'sche Zeitschrift [Danckelmann's Newspaper], 1900.

To Page **122.** Kessler, "Ueber die Aufforstung von Oedlaendereien" [On the Reforestation of Wastelands], *Zeitschrift fuer Forst und Jagd* [Journal for Forestry and Hunting], volume 15, page 427.

To Page **123.** Sprengel, "Eine forstliche Studien-Reise usw." [A Forestry Educational Travel, etc]., Berlin 1879.

To Page **125.** Prince Pueckler, "Andeutungen ueber Landschafts-gaertnerei" [Hints on Landscape Gardening]. (A magnificent book supplied with large picture tables).

To Page **125.** Dr. Zwierlein, "Vom grossen Einfluss der Waldunge-nauf Kultur und Beglueckung der Staaten" [On the Great Influence of the Forests on Culture and Happiness of the States]. (Wurzburg 1806, page 67).

To Page **125.** Cotta, "Die Baumfeldwirtschaft" [The Combination System of Forest and Farm Crops], Dresden 1819, 1st issue.

To Page **126.** Shelterbelts: Mitteilungen aus dem Forstbetrieb im Regierungsbezirk Wiesbaden [Statements out of the Forest Management in the Adminstration District of Wiesbaden]. Put together by Head Forester von Bornstedt for the first General Meeting of the German Forest Society in Wiesbaden in the year 1900.

Chapter Two.

To Page **131.** Natural Subdivision: Weise, "Die Taxation der Privat- und Gemeinde-Foresten" [The Taxation of the Private and Community Forests], Berlin 1883, paragraph 27.

Ritter von Guttenberg, "Die Forstbetriebseinrichtung" [The Forestry Management Establishment], Vienna 1896, page 35.

To Page **131.** Neumeister, "Forsteinrichtung der Zukunft" [Forest Installation of the Future], Dresden 1890.

To Page **134.** Denzin in the *Allgemeine Forst und Jagd Zeitung* 1880, page 126.

To Page **142.** von Guttenberg, "Die Pflege des Schoenen in der

Land- und Forstwirtschaft" [The Care for the Beautiful in Agriculture and Forestry], *Oesterreichische Forst-Zeitung* [Austrian Forest Journal] 1889.

To Page **148.** "Schulze's Fleiss" [Schulze's Diligence], *Forstliche Blaetter* [Forestry Papers] 1888.

Chapter Three.

To Page **149.** Yellowstone, Velhagen & Klasings *Monats-Hefte* [Monthly Journals], VI, issue 8.

To Page **150.** Herrenhausverhandlungen [Negotiations in the Upper Chamber], see "Forstaesthetische Tagesfragen" [Forest-aesthetic questions of Today]. *Zeitung fuer Forst und Jagd* 1898, page 333. "Die Bedeutung des Grunewaldes fuer das Berliner Publikum" [The Relevance of the Grunewald for the Berlin public]: Preuss, "Die Lunge Berlins" [The Lungs of Berlin], *Daheim* [At Home] 1896, Nr.48.

To Page **150.** Luckenurwald [Lucken Primeval Forest]. *Gartenlaube* [Arbor] 1901, Nr.48.

To Page **150.** I owe the information on the Neuenburger Urwald [Neuenburg Primeval Forest] to handwritten statements of the Grand-ducal Forest Supervisor Cropp in Oldenburg.

A picture portfolio with good photographs after paintings by J. Preller and a short text has been published by J. W. Aquistapace in *Varel At The Jade.*

To Page **151.** Selection Forest: Praised in comparison to high forest by Wurm, "Waldgeheimnisse" [Forest Secrets], 2nd edition.

Northwestern German Forest Society June 1884, meeting in Hameln.

To Page **152.** von Guttenberg, as mentioned above.

To Page **152.** Schleiden, "Fuer Baum und Wald" [For Tree and Forest].

To Page **158.** Danckelmann, *Zeitschrift fuer Forst und Jagd* 1881.

To Page **160.** Head Forester Schrember, *Baurs Centralblatt* [Baur's Central Paper] 1877.

To Page **160.** Burckhardt, "Die Waldflora und ihre Wandlungen"

[The Forest Flora and its Changes] ("Aus dem Walde" [Out of the Forest] V), page 104.

Chapter Four.

To Page **166.** Wilbrand, *Allgemeine Forst und Jagd Zeitung* 1893, page 77.

Chapter Five.

To Page **172.** Martin, "Die Folgen der Boden-Reinertragstheorie" [The Consequences of the Theory of the Highest Net Profit of the Soil], Leipzig at Teubner 1894 ff.

To Page **172.** Theory of the highest net profit. The direction, making decisions against the responsibility of the forest aesthetician, has been represented, among others, by Head Forester Bruhm in the Saxonian Forest Society, but soon there was some protest in the Assembly (*Report on the 41st meeting of the Saxonian Forest Society* and *Zeitschrift fuer Forst und Jagd* 1898, page 328).

To Page **173.** Pressler, "Hochwaldideal" [High Forest Ideal], 3rd edition, page 160.

To Page **173.** Neumeister, "Die Forsteinrichtung der Zukunft" [The Forest Institution of the Future], Dresden 1890, pages 1 and 2.

To Page **173-174.** Wilbrand, *Allgemeine Forst und Jagd Zeitung* 1893, pages 119 and 120.

To Page **174.** Wilbrand, as mentioned above, page 121.

To Page **176.** Pfeil, "Kritische Blaetter" [Critical Papers] VII, 2, page 73.

To Page **176.** Beech: Pressler, Rat. F. W. Flugblatt I [Flyer], page 30, and "Hochwaldideal" [High Forest Ideal], 3rd edition, page 160.

To Page **176.** Factory: Pressler, R. F. W. Flugblatt I, page 54.

To Page **177.** Head Forester Poepel in the Wiener *Centralblatt fuer das gesamte Forstwesen* [Vienna Central Paper for the Entire Forestry] 1890, page 493.

To Page **178.** R. Hartig, *Forstliche Naturwissenschaftliche Zeit-schrift* [Forestry Scientific Journal] 1892.

To Page **178.** Hufnagel, "Die Grundzuege der wahren Bestandeswirt-schaft" [The Basics of the True Economical Stand Management], *Vereinsschrift des boehmischen Forst-Vereins* [Society Paper of the Bohemian Forest Society] 1898/99, issue 5.

To Page **179. "Priest"**—words of the State Head Forester von Hagen at the anniversary of the Forestry Academy at Eberswalde (*Zeitschrift fuer Forst und Jagd* 1880, page 416).

Chapter Six.

To Page **181.** G. L. Hartig. Cotta has used this remark for his combined system of forest and field crops; there (Issue 1, page 34) I have found the quotation.

To Page **183.** Forestry Advisor L. Hampel, as mentioned above, (to page 15).

To Page **184.** Boden, Royal Forest Supervisor, "Die Laerche" [The Larch], Hameln and Leipzig 1899.

Chapter Seven.

To Page **189. Postel Thinning Method:** *Allgemeine Forst und Jagd Zeitung* 1893, page 225. Report on the negotiations of the Northwestern German Forest Society at Hannoversch-Muenden 1894, pages 50 and 51. *Zeitschrift fuer Forst und Jagd* 1898, page 672. Treatments of the Bohemian Forest Society 1900 (Society Journal 1900/01, Issues 2 and 3, page 97).

To Page **191.** Ney, Jahresbericht [Annual Report] 1890, page 163.

To Page **191. Selection Method of Thinning.** The modified selection method of thinning, as it is handled here, is described very well by von Bentheim in his "Anregungen zur Fortbildung von Fortwirtschaft und Forstwissenschaft im 20. Jahrhundert" [Ideas for the Continuing Education of Forestry Management and Forestry Science in the 20th Century]", page 57 (Trier 1901). This method, however, according to my opinion, is not contradictory to the Postel Thinning Method, but follows it, beginning with the about 60th year of the stands. The transition is done by pruning. von Salisch: *Erste Durchforstung eines Kiefernbestandes* [First Thinning of a Pine Stand], Zeitschrift fuer Forst und Jagd 1898, page 672.

To Page **193–195.** Prunings: Professor G. Hempel, "Die Aestung des Laubholzes, insbesondere der Eiche" [Pruning of Deciduous Trees, especially of the Oak], Vienna 1895, and Th. Heyer, "Die Vornahme von Aufaestungen in der Oberfoersterei Schiffenberg" [The Way of Prunings in the Forest Administration Schiffenberg], *Allgemeine Forst und Jagd Zeitung* 1901, page 83.

To Page **195.** Parisius, "Leitfaden fuer den Betrieb des praktischen Obstbaues" [Guideline for the System of Practical Fruit Cultivation], Hildesheim 1889. **Against the cettle-pruning** of the fruit tree, also see *Pomologische Monatshefte* [Pomological Monthly Journals] 1901, page 229.

Chapter 8.

To Page **196.** Ebermeyer, "Die gesamte Lehre von der Waldstreu" [The Complete Teaching of the Forest Litter], Chapter 2.

To Page **198.** Homer (see Ebermeyer as mentioned above).

To Page **200.** Czleb, *Jahrbuch des schlesischen Forst Vereins* [Annual Book of the Silesian Forest Society] 1897, page 202.

To Page **201.** Head Forester Tessmann, Burckhardt, "Aus dem Walde" [Out of the Forest], Issue 9.

To Page **202.** Rothe, "Ethik und Aesthetik im Waidwerk" [Ethics and Aesthetics in Hunting] (Neudamm 1901) only touches upon aesthetics of hunting.

Chapter Nine.

To Page **205. Branching of the water courses:** von Duecker, "Zur Frage der Wasserpflege" [On the Issue of Water Management and Care], *Zeitschrift fuer Forst und Jagd* XIII.

PART II, Section B.

Chapter One.

To Page **214.** Weise, "Leitfaden fuer den Waldbau" [Guideline for Forestry], Berlin 1894, pages 104 and 105.

To Page **215.** Kraft, "Zur Aesthetik der Park- und Waldwirtschaft"

[On the Aesthetics of Park and Forest Management], Zeitschrift fuer Forst und Jagdwesen [Journal on Forestry and Hunting] 1895, page 395.

To Page **216. Upper Chamber,** *Zeitung fuer Forst und Jagd,* June 1891.

To Page **216. "Review of the First Edition,"** *Wiener Zentralblatt fuer das gesamte Forstwesen* [Vienna Central Paper for the Entire Forestry] 1885.

To Page **217. Landscape gardener.** The Grandducal Oldenburg Head Forest Supervisor von Heimburg has done excellent work in regard of forest aesthetics and as a landscape gardener. I mention as being his creations the Game Park at Hankhausen, the open landscape near Eutin, the park near Gueldenstein. The story of his life is in *Zeitschrift fuer Forst und Jagd* 1892, page 577.

To Page **219.** Wilbrand, *Allgemeine Forst und Jagd Zeitung* 1893, page 73.

Chapter Two.

To Page **220. "Voluptar-Forest,"** *Oesterreichische Forst und Jagd Zeitung* 1898, (May 20), No. 20.

To Page **221.** Eisenach, Dr. Stoetzer, "Die Eisenacher Forsten" [The Forests of Eisenach], Eisenach 1900.

To Page **223.** Leuer, as mentioned above (compare with page 15).

Chapter Three.

To Page **225. To the title:** I mention as works, particularly to be noticed for **open landscapes: Pueckler-Muskau,** "Andeutungen ueber Landschaftsgaertnerei" [Hints on Landscape Gardening], Stuttgart 1834; G. Mayer, *der schoenen Gartenkunst* [Textbook of the Beautiful Art of Gardening]; Petzold, "Die Landschafts-gaertnerei" [The Landscape Gardening].

Chapter Four.

To Page **243.** Neumeister, "Zur Schonung der Waldbaeume" [On the Saving of Forest Trees], *Dresden Advertiser* 1892.

To Page **245.** "Spectrum of colors" Hampel, as mentioned above.

Chapter Five.

To Page **246.** Most gardening books contain information on **allees.** As special writings I mention: E. Petzold, "Die Anpflanzung und Behandlung von Alleebaeumen" [The Planting and Maintenance of Allee Trees], Berlin 1878. Fintelmann: *Ueber Baumpflanzungen in den Staedten* [On Tree Plantings in the Cities], Breslau 1877. This little book by Fintelmann is rather indispensable for forest officials in community administrations, who have to supervise urban tree plantings; other people will read it with interest, too.

To Page **254. Fruit trees.** In the XIII. Meeting of the German Pomologen [Apple Scientists] in Breslau, a listing of fruit species suitable for street planting has been provided. (Annual Report, pages 416 and 417).

To Page **256. Telegraph wires.** In the "Pomologische Monatshefte" [**Pomological Monthly Journal**] 1901, page 92, is stated with the hint to the "Elektrischer Anzeiger [Electrical Advertiser]" Nr.87, that the Hackethaldraht-Gesellschaft m.b.H. isolates wires according to a patented method so that they are able to be run through tree crowns without the amount of current being diminished by the touch of the branches.

Chapter Six.

To Page **259.** The supplements to *der Allgemeinen Forst und Jagd-Zeitung* contain very rich listings of literature in respect to **old and odd trees,** as well the **"Merkbuecher"** [notebooks]. (See the note to Pages 40 and 67.) The "Baumalbum der Schweiz" [Tree Album of Switzerland] includes pictures from photographs of magnificent trees, in very large format, Bern 1900.

To Page **260.** R. Hartig, "Zersetzungserscheinungen des Holzes" [Decomposition Phenomena of the Wood], Berlin 1878, page 149.

To Page **261-263.** Fickert, Royal Head Forester in Werder on the island of Ruegen, lately Forest Supervisor in Altruppin.

To Page **264-266. Klipstein-Oak:** According to written information of the Grandducal Ministerial Director Wilbrand, and in *Allgemeine Forst und Jagd-Zeitung* 1893, page 79.

Chapter Seven.

To Page **269. Hampel,** as mentioned above.

To Pages **271. Bratranek,** "Beitraege zu einer Aesthetik der Pflanzenwelt" [Contributions to an Aesthetic of the Plant World], Leipzig 1853, pages 409 and 410.

To Page **272. Forest officials' farms;** decree from March 5, 1898, Danckelmann Jahrbuch 1891, page 135.

To Page **274. Luetetsburger Firs. Burckhardt,** "Aus dem Walde" [Out of the Forest] I and VIII.

To Page **274.** R. Hartig, *Forstlich-naturwissenschaftliche Zeit-schrift* [Forestry Scientific Journal] 1892, issues 11 and 12.

To Page **275-276. Varieties:** von Fischbach has recommend several times to use seedlings of the well-known tree species, horticulturally propagated by grafting, for forest embellishment. *Allgemeine Forst und Jagd-Zeitung* 1848, page 325, and 1861, page 88. *Wiener Centralblatt fuer das gesamte Forstwesen* [Vienna Central Paper for the Entire Forestry] 1893. The pyramid poplar is also supposed to be grown from seed, in order to get trees that stay healthy for a long time. (Dr. Bonhausen, *Allgemeine Forst und Jagd-Zeitung* 1881, page 297.)

Chapter Eight.

To Page **281.** Burckhardt, "Aus dem Walde" [Out of the Forest] X, 24.

To Page **282.** von Heimburg, at that time major-domo in Oldenburg.

Chapter Nine.

To Page **285.** Maechtig, Municipal Garden Director in Berlin, has also created the Viktoria-Park with the area of rocks on the Kreuzberg (*Gartenflora* [Garden Flora], Berlin 1894, page 263).

To Page **289.** Wocke, "Die Alpenpflanzen in der Gartenkultur usw. [The Plants of the Alps in Garden Cultivation, etc.]," Berlin 1898.

Chapter Ten

To Page **290. Ruins,** "Burgwart" [Castle Keeper], *Zeitschrift fuer*

Burgenkunde und mittelalterliche Baukunst [Journal for the Science of Castles and Medieval Architecture], Organ der Ver-einigung zur Erhaltung deutscher Burgen [Publication of the Society for the Rescue of German castles], IIIrd year, (publisher Franz Ebhardt & Co., founded 1867, Berlin W, Schaperstr. 5)

Chapter Eleven

To Page **296.** Pfeil, Hey, *ein Erzieher des deutschen Waldes* [A Grower of the German Forest], Halberstadt 1891, pages 8 and 37.

To Page **299.** Wilbrand, *Allgemeine Forst und Jagd-Zeitung* 1893, page 118.

G. F. Schwarz, *Forest Trees and Forest Scenery,* illustrated, New York, The Grafton Press 1901. This book, rich with observations and nicely made, was given to me by the kindness of its author. Unfortunately, I can only mention it here, because the state of printing *Forest Aesthetics* is already far advanced.

Notes Concerning The History and Literature of *Forest Aesthetics*

[Notes and References Used in Forest Aesthetics, First Edition, by von Salisch in 1885, Berlin]

(Note: References to page numbers do not match the translated text)

[Preface]

1) It is said that as early as 50 years ago the first German *Forest-Aesthetics* was planned to be written. In the year 1830 (page 543) of the *Allgemeine Forst- und Jagdzeitung* [General Forest And Hunting Journal], Baron von der Borch (Bavarian and former Bayreuth Head Forester) was looking for a solid publisher for his "Aesthetics in the Forest," which

which he wanted to publish supplemented with pictures. But he may not have been able to find a Julius Springer, willing to make sacrifices, which is very regrettable, for von der Borch possessed most of the qualities which a forest aesthetics writer should unite in his person. Already for 40 years, the understanding of caring for the forest's beauty had been initiated in him by strong impressions in his youth, as he himself reports in the publication mentioned above, and he also had already showed himself capable of matching his subject in a short essay (published in the Sylvan of 1824). We can find the story of his life, written in a not very benevolent style (unfortunately he has sinned by writing bad verse), in Bernard's *Forstgeschichte* [Forest History], (II. page 402).

2) The same explanation is valid also particularly for the notes following here.

3) Before this summarizing book on the subject, I have written several short essays. They can be found in the *Zeitschrift fuer Forst und Jagdwesen* [Journal for Forestry and Hunting] Volumes VIII, X, XIII, and in the *Jahrbuecher des Schlesischen Forstvereins* [Yearbooks of the Silesian Forestry Society] for the years 1877, 1878, 1879, 1880, 1882, and in the *Allgemeine Forst und Jagdzeitung* [General Journal for Forestry and Hunting] of the year 1879.

"The first step is always the most difficult" says the proverb. For the first steps of my writings on forest aesthetics, I owe much to Head Forester von Bornstedt, now in Kassel, especially ideas in the field of beauty care in the forest, while Consistorialrath Prof. Dr. Kleinert in Berlin has helped, by correspondence, with the emergence of the first draft. Later the president of our Schlesischer Forstverein [Silesian Forestry Society], Forestry Commissioner Dr. A. Tramniss, and several respected members of this Society (among others von Kujawa, Lignitz, and Bosfeldt) have encouraged me in a very friendly way to continue with the work.

[Introduction]

4) To page 1. The *Journal of Forestry*, published in London, has set for itself the task of cultivating also the aesthetic aspect of forestry. In doing so, it published (in the issue of May, 1877) an estimable essay written by a clerical gentleman, the Reverend J. G. Macvicar. There I found **for the first time the art of forestry set beside architecture.** Macvicar writes: "It can truly be said that the forester takes a position in art which is equal to the one of the architect. However, a ballot might have a result favoring the latter, if the architect were not able to bring together a lot of ugliness and shapelessness, which nature will never allow the forester to create."

Unfortunately, I cannot tell whether the British colleagues have been working industriously on the expansion of the young branch of our subject in the meantime, because I have not followed the Journal, not knowing the English language.

5) To page 3. According to this, a knowledge of the literature of the fine art of gardening will at least be desirable for the forester. As recommended writings I name: Pueckler-Muskau: *Andeutungen ueber Landschaftsgaertnerei* [Hints on Landscape Gardening], Stuttgart 1834; G. Mayer: *Lehrbuch der schoenen Gartenkunst* [Textbook on the Fine Art of Gardening]; E. Petzold: *Die Landschaftsgaertnerei* [Landscape Gardening], Leipzig 1862. The book by Pueckler unfortunately is rather costly, due to the beautiful large color plates with which it has been supplied, and the German does not like to spend much money on books, therefore this pioneering book is lacking even in most of the larger libraries. I myself owe the knowledge of this book only to the graciousness of the Administration of the Royal Saxonian Court Library.

Whoever, prohibited by lack of time or otherwise, does not wish to study one of the mentioned larger books, might want to read a very inspiring **small booklet** by G. Petzold: *Fuerst Hermann von Pueckler-Muskau in seinem Wirken in Muskau und Branitz, sowie in seiner Bedeutung fuer die bildende Gartenkunst Deutschlands* [Prince Hermann von Pueckler-Muskau in His Works in Muskau and Branitz, as Well as His Importance To the Fine Art of Gardening in Germany], (Erlangen 1874 by F. Enke.)

6) To page 4. So for example B. G. Heyer wrote in the *Supplement zur Allgemeinen Forst und Jagd-Zeitung* [Supplement to the General Forestry and Hunting Journal] X 26: "If a private person manages his forest in a park-like way and keeps all the picturesquely beautiful but economically mature trees in it, he cannot be prohibited from doing so, for he has to account for his property and its management only to himself. But it is different with federal forests. The revenues coming out of these serve to decrease the tax burdens of all citizens, so even the poorest one is interested in the federal forests bringing the highest net profits, for then he will be required to pay less taxes."

7) To page 5. Forestry Inspector Schaal: *Allgemeine Forst und Jagd-Zeitung* [General Forestry and Hunting Journal] 1879, page 260. With much more emphasis also: Head Forester Guse: *Forstliche Blaetter* [Forestry Information] 1876 and 1877. Quite recently: Head Forester Neihardt, *Forstwissenschaftliches Central Blatt* [Forestry Central Paper], 1884, page 294, and 1885, page 144.

8) To page 5. The comment on the 2nd edition of Koenig: Waldpflege [Forest Care], in the *Allgemeine Forst und Jagd-Zeitung* [General Forestry and Hunting Journal] 1876, is very characteristic of the behavior of the critics. This book, in its revised shape by Grebe, includes nearly 500 pages, and of these, less than five deal with "enhancement of the forests," and how does the overly strict Mr. Critic act against these in his essay, filling 19 columns? Obviously he believes that he has to fight an abundance of unproductive attitudes, therefore he joins in nearly **two lines** the discussed section "with pleasure," but then he immediately protects the rights for a management in the sense of the highest net revenue. The critic is silent on the other subjects, and he does not object that that chapter had been written and is still standing there today without making use of Pueckler's theories concerning this subject, which had already been introduced into the literature at that time.

[Chapter 1. Principles of General Aesthetics]

9) To page 8. I would like to recommend, to someone who has enough time to read **larger** aesthetic writings, first of all the following:

Jungmann: *Aesthetik* [Aesthetics], Freiburg im Breisgau 1884. Unfortunately I happened to find this heavy volume just a few days ago, so that I was not able to make use of its contents. The book, written very clearly and understandably, can be read easily, so that one does not need to worry due to its size. However, the field of beauty in nature, which is particularly important to us, has been ignored by the author, therefore one would, besides Jungmann, also need the following one:

Berthold: *Das Naturschoene* [Beauty in Nature], Freiburg im Breisgau, 1875. This book, too, is pleasant to read and easy to understand, but it is smaller than the one mentioned above, and is nicely supplied with wood engravings.

Both Jungmann and Berthold write with a specifically Christian point of view, the first one with a definitely Catholic one. Whoever does not like that might want to read as an introduction: Lemke: *Populaere Aesthetik* [Popular Aesthetics], Leipzig 1865.

10) To page 9. Oerstedt: *Der Geist in der Natur* [The Mind {Spirit} in Nature], Leipzig 1854, deals with only some of the fields, but thoroughly with those ones—mostly in the form of vivid conversations.—His contemplation on the **seemingly unsightly in nature** are very appealing.

11) To page 9. I think the basis for this attitude is Kant's "Kritik der aesthetischen Urteilskraft" [Criticism on The Aesthetic Faculty of Judgement]. I say **"I think,"** because, to be honest, my faculty of perception is not sufficient to understand Kant.

12) To page 11. Jungmann's book, already mentioned above, deals in detail, and very convincingly, with this particular field of aesthetics.

13) To page 15. This has been proved in detail by Zeisig: "Aesthetische Forschungen" [Aesthetic Research], Frankfurt am Main, 1885.

14) To page 22. Fechner: *Vorschule der Aesthetik* [Pre- study of Aesthetics], Leipzig 1876, treats "the principle of aesthetic association" with particular detail.

[Chapter Two. Theory of Colors In The Landscape]

15) To page 25. F. Th. Bischer has acted as a pioneer for the theory of Beauty in Nature (see his *Aesthetik* [Aesthetics], zweiter Teil [second part], Leipzig 1847.) Also the writings of Oerstedt and Berthold, already mentioned above, need to be mentioned again at this place.

Out of the large number of recent books and booklets which talk about single fields of beauty in nature, suitable for a larger audience,

I would like to name: Masius: *Naturstudien* [Nature Studies], which was published in several editions and is justifiedly widely appreciated. **Additionally, Leopold Schmerz: *Naturgeschichtliche Charakterbilder* [Character Studies of Natural History],** Leipzig 1879. Schmerz, too, presents some of his observations in appealing form, but not all of his chapters are equal among themselves.

16) To page 25. Altum: ***Der Vogel und sein Leben*** [The Bird And Its Life], Muenster 1875, is, according to its objective, not particularly dedicated to the contemplation of the beauties in the bird's world, but nevertheless, one can find here rich knowledge concerning this aspect and also more personal inspiration to knowledgeably looking (and listening!).

17) To page 25. Basically, I used as a guide Helmholtz (*Populaere wissenschaftliche Vortraege* [Popular scientific lectures], Braunschweig 1876, Volume II). At his time, E. Petzold's theory (*zur Farbenlehre der Landschaft* [On the Theory of Color In the Landscape], Jena 1853) was still based on Goethe's theory of the colors; nevertheless the practical hints, which Petzold gives, are still valuable today.

The reader is going to find detailed and up-to-date instruction in the nicely decorated book: von Betzold, "die Farbenlehre im Hinblick auf die Kunst und Kunstgewerbe" [The Theory of Colors In Respect to Art And Arts And Crafts], Braunschweig 1874. Every reader should get this book for his wife, if he has one, and if she has the time for colorful needlework and reading books.

18) To page 36. Rossmaessler: *Der Wald* [The Forest], page 23 of the first edition. This book has already been published in its third edition, which has been revised by Willkomm after the author's death.

[Chapter Three. Tree Species]

19) To page 55. William Gilpin was at that time Canon at Salisbury and Minister in Boldre in New Forest. In 1791 he wrote two volumes with the title: "Bemerkungen ueber Waldscenen und Waldansichten und ihre malerischen Schoenheiten, von Scenen des New-Forest in Hampshire hergenommen." [Remarks on Forest Scenes and Views And Their Picturesque Beauties, Deducted From Scenes of the New Forest in Hampshire]. (The German Translation was published in Leipzig in 1800).

This is a very valuable book which later was quoted very frequently, among others by Masius. The *Allgemeine Forst und Jagd-Zeitung* [General Forestry And Hunting Journal] from the year 1831 published extensive excerpts of this book.

20) To page 57. *Lehrbuch fuer Foerster* [Textbook For Foresters], 5th edition. Also, an artist's eye makes the difference, as one can see in the excellent pictures of both oak species by J. Fischbach in the marvelous book "J. Fischbach und H.MASIUS: Deutscher Wald in Wort und Bild" [The German Forest in Words and Pictures], (Muenchen, Friedrich Bruckmann).

21) To page 60. Bratranek: *Beitraege zu einer Aesthetik der Pflanzenwelt* [Contributions to an Aesthetic of the Plant's World], Leipzig 1853.

22) To page 62. Hermann Jaeger, Grandducal Saxonian Court Gardener in Eisenach: **"Deutsche Baeume und Waelder.** Popular-aesthetische Darstellungen aus der Natur und Naturgeschichte und Geographie der Baumwelt" [German Trees And Forests. Popular Aesthetic Presentations from Nature and Natural History and Geography of the Trees' World]. Leipzig, Karl Scholtze.

The book offers more than the title "popular-aesthetic" promises. It is interesting also for the educated forester.

23) To page 67. Rossmaessler: *Der Wald* [The Forest], first edition, page 262.

24) To page 70. Literally quoted from Bischer: *Aesthetik, Die Lehre vom Naturschoenen* [Aesthetics, The Theory of Beauty in Nature]. p. 272.

25) To page 74. *Darmstaedter historische Kleinigkeiten* [Darmstadt Historical Minor Points], Darmstadt 1879.

26) To page 74. *Forstliche Blaetter* [Forestry Information] 1878, page 85.

[Chapter Four. Voice And Fragrance of The Forest]

27) To page 81. Such an examination, which would not be allowed to stop at non-living nature, finds valuable material in Altum: *Der Vogel* *und sein Leben* [The Bird And Its Life], also **some material in Schleiden: "Studien" [Studies]. Leipzig 1857.**

[Part Two]
[Section A. The Forest Establishment]

[Chapter One. The Determination of The Most Useful Type of Soil Use]

28) To page 85. The social meaning of the care for beauty in the forest has been duly acknowledged also by **forestry,** for example by Head Forester Bosfeldt in the *Yearbook of the Schlesischen Forstvereins* [Silesian Forest Society] for 1872:

"Ueber die soziale Frage und den Materialismus in der Forstwirtschaft" [On the Social Question And the Materialism in Forestry]. Also a longer talk by Dr. Heinrich Contzen on "Die ethische and aesthetische Bedeutung des Waldes" [The Ethical And Aesthetic Meaning of the Forest], printed in his "Forstliche Zeitfragen" [Forestry Contemporary Questions]. (Leipzig 1870, M. G. Priber.)

Also Bernhardt has warmly emphasized the social meaning of the forest, for example in the *Zeitschrift fuer Jagd- und Forstwesen* [Journal for Hunting and Forestry] (VI 343): "The human does not live on bread alone—even less by gold alone—, but there is something, which cannot be calculated or be expressed with figures, something which does not yield interest measured in percentages, but still belongs to a complete human and which stands much higher than all the gold-making art of modern times; that is the deeply-moral consciousness of nature's harmonic beauty with which we are not allowed to interfere destructively without being punished, of the **duties which we all have to fulfill towards our fellow men, of their right to physical and mental health and a humane existence."**

Quite recently (1885, Vol. I), the "Forstwissenschaftliches Centralblatt" [Forestry Central Paper] has published an article, written with much empathy by its editor (Prof. Dr. Franz Baur), which further discusses the same idea and reaches the demand that it has to be said to our "studying youth," that the forester also **"is not allowed to unnecessarily restrict the rights of the social politicians."**

In this essay, which contains many interesting ideas, for example, the proof of how much all the arts, music included, owe to the forest; the hint on the **"studying youth"** has been particularly valuable to me.

There are characters which become more and more rigid in later life, and which depart so far from the attitudes of Arndt, Riehl, Koenig or Burckhardt that they believe not to know any more important duty than to block the public from the forest as much as possible. If they had power over the path which the poet SCHILLER was going in his "Spaziergang" [Walk]:

> *"Only secretly spare light is shining through the branches'*
> *Loose grid, and the sky's blue looks in laughingly.*
> *But suddenly the veil tears. The opened-up forest*
> *Surprisingly returns me to the day`s blinding glare."*

they certainly would come to the conclusion that the path has to be removed as being not necessary for the public traffic. It is to be wished that with the maturing forestry youth such hidebound attitudes are prevented rather thoroughly.

29) To page 85. Der Waechter [The Guardian], Volume II (1815) "Ein Wort ueber die Erhaltung der Forsten und der Bauern im Sinne einer hoeheren, d.h. menschlichen Gesetzgebung" [A Word On The Maintenance Of The Forests And of The Farmers In The Sense Of a Higher, That Is Humane Legislation]. Pages 399ff and 387.

30) To page 86. Riehl: *Land und Leute* [People And Country], Stuttgart.

31) To page 90. Bernhardt: *Ueber Baumkultus und Waldsagen* [On Tree Cult And Forest Tales], Eberswalde 1873.

32) To page 90. This miraculous indifference becomes explicable only by the fact that the clearings of the small woods happen **little by little.**—Else, the otherwise so powerful voice of "public opinion" would probably not remain so still like a mouse. We Germans are so vain about us loving the forest and the trees so much!

How painfully felt it was already in presumably rougher times, if the destruction of the last remnants of the old forest happened **all of a sudden;** for that I found in an hour of leisure a fairly touching proof in the old **Josephus.** Here I quote the passage as I copied it out of a German translation from the year 1574.

In the 6th Book, 1st Chapter on the **Jewish War** (siege and destruction of Jerusalem by Titus) it is said: "Although the Romans, with the planned construction/ and also the transport of the timber/ had great labor to do/ nevertheless the truss of the bridge was erected by them within twenty-one days/ also **all the forests and woods eleven miles around the city were cut down/ therefore the Jewish country had become rather barren and unsightly/** which before had been adorned with green forests and nice gardens. But after all the trees had been cut down everywhere/ it looked like a wilderness and was so devastated/ that no foreigner/ if he had seen before the magnificent country and the tremendous sites/ and now looked at the starvation/ could refrain from crying and sighing **because of the change that had happened. For the war had taken away all the adornments and all the beauty/** and if someone/ who before had known the country well / came there unexpectedly/ he surely would not have known the place anymore/ but would have needed to ask for the way like a stranger."

This is probably the oldest **forest aesthetic** note, if one does not want to consider paradise also as a large **forest** garden.

33) To page 91. Cotta: *Die Baumfeldwirthschaft* [The Combination System of Forest And Farm Crops], Dresden 1819, vol. I.

34) To page 93. At the moment, the Humboldt-Hain [Humboldt-Grove] and the Friedrichshain [Friedrich-Grove] are under the direction of Roennenkamp.

35) To page 96. The "Buchwald" [Beech forest] near Trebnitz, a forest plot belonging to the Royal Head Forest District Kuhbrueck, is managed as luxury forest in a very proper way.

We Germans think we have already done a lot if we take some 100 ha out of economic purposes. Across the ocean, people have roused themselves to make quite different decisions. The Congress of the North American United States has set aside by law an area of the size of a province as the so-called **National Park of Yellowstone** in the territory of Wyoming, south of the North Pacific Railway. Unfortunately I do not really know what to understand by a "National Park." Because I do not know the English language, I also couldn't find out more details about this.

36) To page 96. It is not necessary for the purse of the forest owner **himself** to pay for the little expenses for enhancement purposes. At the time when I was forester at Altenplathow, there was an **Association for Embellishment in Genthin,** whose committee was as skilful in the raising of money as in using the collected funding for the small city forest.

There are associations or societies for the mountains, like the Alps, the Vogesen, the Riesengebirge [Sudenten Mountains], whose activity often touches the tasks of the art of forestry in the luxury forest. It seems that an association for embellishment near the town of Bruenn **proceeds especially successfully with reforestation.**

[Chapter Two. Road System And Management Units]

37) To page 101. Muehlhausen: *Wegenetz* [Road System]. Frankfurt am Main 1876.

38) To page 102. Dr. Carl Grebe: *Die Betriebs- und Ertrags-regulierung* [The Regulation of Management System And Yield]. WEIMAR 1867. page 40. The same source also briefly mentions the separation of parts of the forest for certain (aesthetic) purposes.

39) To page 108. *Allgemeine Forst- und Jagdzeitung* [General Forest and Hunting Journal] 1880, page 126.

40) To page 112. This chapter was written a long time ago (part has already been published in the *Zeitung fuer Forst- und Jagdwesen* [Journal For Forestry And Hunting] in the year 1881); in the meantime, Weise's "Taxation der Privat- und Gemeindeforsten" [Taxation of Private And Municipal Forests] (Berlin 1883) has been published, and after looking through this valuable booklet I thought already that the "regular division" was an attitude left behind, which is no longer worth fighting against. Therefore I was about to rewrite the manuscript again and to shorten it considerably. But just now I learnt from a map I received, the map of a brand new forest establishment stemming from well-known foresters, how eager, I would nearly say blind, supporters of the regular division still are.—The mentioned map shows that among 33 compartments, 31 are edge-compartments, so, according to Adam Riese, there are only two left which have entirely the desired size and shape. For the sake of these two equal compartments, all the rest has not only been cut into partly disgusting tips, but also long firebreaks have been designed and cut open quite close beside good old roads (which continuously have to be maintained!)

41) To page 115. At the same place, page 65.

[Chapter Three. The Silvicultural Systems]

42) To page 120. Dr. H. Martin: *Wegenetz, Eintheilung und Wirthschaftsplan in Gebirgsforsten* [Road System, Division And Management Plan in Mountain Forests]. Muenden 1882.

43) To page 122. See the essay of Head Forester Danckelmann on "Kiefern-Unterbaubetrieb" [Pine-Understory-Management], *Zeitschrift fuer Forst- und Jagdwesen* [Journal For Forestry And Hunting] 1881. From this essay, which is very valuable for forest aesthetic efforts, I quote the remark: "As already mentioned, **beeches and hornbeam poles and minor tree species were partly kept as standards individually and in groups** in the regeneration of the mixed pine stands grown up from selection forest. After having become mature, these have created for themselves all around them, a dense, ground covering understory under the light pine crowns by repeated seed-fall. The stand scenes emerging from that have become the model for the pine-understory-management established here."

For my part I can add that also here in Postel, single older beeches, surrounded by rich seed sprouts, very much adorn several pine pole stands, a forest part which otherwise might be barren and boring.

44) To page 125. The Nordwestdeutsche Forstverein [North West German Forest Society], during its first travel meeting at Hameln (22nd to 24th of June, 1884) has put a forest aesthetic subject onto the agenda whose title was: **"Is it desired to choose at appropriate sites silvicultural systems which permit a unification of aesthetic aspects with the economical ones, and which of the silvicultural systems can be considered for this purpose?"** The Province Head Forester Quaetfaslem, Head Forester Wallmann, Count Knyphausen, have spoken on this subject and fundamentally made clear the decisive aspects concerning the "luxury forest." As a silvicultural system preferably suited for this purpose, the **plenter** [selection] **management** was recommended.

I took this message out of the *Zeitung fuer Forst- und Jagdwesen* [Journal for Forestry And Hunting] 1884 page 495. I have not yet received the *Annual Report of the* **Nordwestdeutschen Forstverein [North West German Forestry Society].**

[Chapter Four. The Choice of The Tree Species]

45) To page 130. How a castle ruin has to be treated is told by GOETHE very much in detail in a "novelle," which is found in the "Unterhaltungen deutscher Ausgewanderter" [Conversations of German Emigrants]. The same subject and the meaning of architecture in landscape in general is dealt with very aptly and in an attractive form in Utis: "Der falsche Baurath, eine Novelle fuer Kunst- und Altertumsfreunde" [The False Head of The Planning Department, a Novellete For Friends of Art and Antiquity]. Frankfurt am Main 1877.

[Chapter Five. The Rotation Period]

46) To page 137. Meyer: *Lehrbuch der schoenen Gartenkunst* [Textbook of The Fine Art of Gardening].

47) To page 143. Cotta has used this place for his combination system of trees and fieldcrops. There I have found it (Booklet I page 34).

It has remained doubtful to me **whether also Pfeil may have had forest aesthetic feelings.** It seems that he had the opinion that there is no space left for the care of beauty beside the striving for the satisfaction of **material** needs, and therefore he probably has not even tried to solve a task that he thought of as impossible from the beginning. But it may be that I am wrong about this. My assumption is based only on one single place (*Kritische Blaetter* [Critical Papers] Volume VII, Booklet II, page 73), which says: "The material need allows less and less space to the sense of beauty in forestry. First the magnificent large old trees disappear, then the single picturesque tree groups, and at last the monotonous gray dead pine drives out the friendly lively deciduous trees. The same appearance which takes place in the change between the poetry of the hunter's life and the dull eking out of an existence in the factory, where the human is only regarded as a part of machine, returns everywhere. **One may bewail that, but one cannot change it!"**

[Chapter Six. Taking Care of The Stand]

48) To page 147. In any case, Prof. Dr. Baur seems not to have found such a procedure mentioned in our literature. See his essay **"On the history of thinnings"** in the *Centralblatt* [Central Paper] 1882. However, in France the same attempt has also been made recently as I can see from a note in Saalborn's "Jahresbericht" [Annual Report] (Fourth Year page 58).

[Chapter Eight. Secondary Uses]

49) To page 151. Ebermayer: *Die gesammelte Lehre der Waldstreu* [The Collected Teaching of Forest Litter], Chapter 2. By the way, the sentence was quoted from Ebermayer, who had quoted it from Berthold Sigismund.

50) Ebermeyer, at the same place.

51) To page 154. Burckhardt: *Aus dem Walde* [Out of The Forest], Booklet 9.

[Chapter Nine. Forestry Calculation]

52) To page 157. For details see Borggreve: *Forstwissenschaftliche Tagesfragen* [Contemporary Questions in Forestry]. Bonn 1878.

53) To page 158. Borggreve, at the same place. Page 43 and 44 the notation.

54) To page 160. Pressler: *Hochwaldideal* [Ideal High Forest], (3rd edition, page 160).

55) To page 160. *Nationales Forst-Wesen Flugblatt* [National Forestry Flyer], I page 24.

56) To page 160. *Nat. Forst-Wesen Flugblatt* [Nat. Forestry Flyer], I page 30 and "Hochwaldideal" [Ideal High Forest], 3rd edition page 163.

57) To page 161. *Nat. Forst-Wesen Flugblatt* [Nat. Forestry Flyer], I page 54.

58) To page 162. Words of Superior State Forester von Hagen at the anniversary of the Forestry Academy at Eberswalde (*Zeitschrift fuer Forst- und Jagdwesen* [Journal For Forestry and Hunting], 1880 page 416).

[Section B. Taking Care of The Forest]

59) To page 163. In the field of the aesthetic care for the forest, forestry literature already has two valuable discourses written by Burckhardt (one chapter of his "Saeen und Pflanzen" [Sowing and Planting]) and Koenig (one chapter of his "Waldpflege" [Forest Care]).

[Chapter One. Park-like Landscapes]

60) To page 164. Up to now very little has been written **by foresters** on the topic of park-like landscapes; I hardly know anything to list here belonging to that topic besides the **essay by Borch** in the *Sylvan,* already mentioned in Note 2). However the writings of landscape gardeners, mentioned several times, contain more or less detailed instructions for the establishment and care of park-like landscapes. Only after finishing this chapter I received an estimable new booklet: Abel: *Die Baumpflanzungen* [Tree Plantings], Vienna 1882. It contains **numerous** pictures of landscapes, first in their natural state, then drawn after the establishment of park-like landscape, which are very informative.

Whoever has read his Goethe with attention has probably found also much there which belongs to the topic. In the "Wahlverwandschaften" [The Elective Affinities], the fates of the four persons, who attract our attention so much, are as closely connected with the enhancement of a wide landscape as are the fates of the Greeks with the battle against Troy, in the *Iliad.* One might want to remember: Eduard has taken care of the garden, quite improperly, for he should have left this to his wife ("the garden is the house extended into outdoors," says Prince Pueckler, therefore the mood is allowed to be at work there). At the same time, Charlotte is working on the park-like landscape. Now the Captain arrives and brings home to both of them that Charlotte is not capable of this task. Charlotte sees that he is right. Being somewhat offended in the beginning, she soon works together with him on a larger scale. In the course of this, she falls in love with the Captain. Now I skip the remodelings in the park to which some memorable events are connected, only to mention that it is Ottilie alone, who, with the right instinct, discovers the only suitable site for the summer house to be erected in the park-like landscape, after which Eduard falls in love with her even more than before.

Who cannot or does not want to realize that such park-like landscapes are really a serious and important thing, might want to read that again in the "Wahlverwandschaften" [The Elective Affinities].

In "Wilhelm Meister's Wanderjahre" [Wilhelm Meister's Travel Years], there are also quite a few valuable remarks on that. So the successful park-like landscapes of the "Senior Civil Servant" are described in detail in the novella "Wer ist der Verraeter?" [Who is The Traitor?], which is put in there (eighth chapter of the first book of the "Wanderjahre" [Travel Years].

61) To page 170. Dr. Maximilian Ritter von Proskowetz has firmly pointed out in "Oesterreichisches Landwirtschaftliches Wochenblatt" [Austrian Agricultural Weekly Journal], Vienna 1880, **how much the landscapes, the people and cattle have lost their picturesque attraction and characteristic properties by the modern way of farming.** "The painter does not show up any more in our industrialized, highly civilized areas, which now offer to him only social, starkly realistic 'material,' and he needs to travel, driven back like the Indian is driven out of his hunting-grounds by the incoming roaring stream of the civilization era which cannot be stopped."

Proskowetz recommends **planting of trees** at places where they will not do damage to agriculture; besides he urges giving a picturesque silhouette to buildings (castles) by **rich roof and gable constructions.**

[Chapter Two. Meadows, Water and Fields in The Forest]

62) To page 173. von Duecker: *Zur Frage der Wasserpflege* [On The Question of Water Care]. *Zeitschrift fuer Forst- und Jagdwesen* [Journal For Forestry And Hunting], XIII.

63) To page 177. Schulz-Lupitz: *Die Kali-Duengung auf leichtem Boden* [Fertilization With Potash On Sandy Soils], Berlin 1882.

[Chapter Three. Establishment of The Roads, Cross-Roads, Roadsigns]

64) To page 182. G. L. Hartig: *Anweisung zur Taxation u.s.w.* [Instruction For Taxation etc.] (2nd edition) shows a **nice copperplate** on the accompanying forest map which one should take as a model in this respect.

See also FISCHER, Grandducal Baden Head Forester: *Album forst- und waidmaennischer Denkmale* [Album of Forestry and Hunting Monuments], Karlsruhe at Malsch and Vogel.

[Chapter Four. Planting of Trees Along Roads and Firebreaks]

65) To page 188. The textbooks of landscape gardening, already mentioned above, also the book by Abel, contain more or less detailed instructions on allee plantings. But there are also monographs on the same subject: E. Petzold: *Die Anpflanzung und Behandlung von Alleebaeumen* [The Planting And Maintenance of Alley Trees], Berlin 1878.

Fintelmann: *Ueber Baumpflanzungen in den Staedten* [On Tree Plantings in Cities], Breslau 1877. Fintelmann's booklet is rather indispensable for municipal forestry civil servants who are responsible for urban tree plantings, but other persons will read it with interest as well.

66) To page 194. Oberdieck: *Deutschlands beste Obstsorten* [Germany's Best Fruit Tree Species], Leipzig 1881.

One can find very detailed instructions, based on very rich experience, concerning the **planting and maintenance of fruit trees along roads** in the 1883 issue of the *Verhandlungen der Schlesischen Gesellschaft fuer vaterlaendische Cultur* (Section fuer Obst- und Gartenbau) [Treatises of the Silesian Society for National Cultivation {Section for Fruit and Vegetable Horticulture}], written by State Building Inspector Sutter.

[Chapter Five. Enhancement of The Forest Stands by Care of Venerable Trees]

67) To page 197. The book by Meyer, already mentioned several times, contains an excellent **Geschichte der Gartenkunst** [History of the Art of Gardening].

68) To page 201. R. Hartig: *Zersetzungserscheinungen des Holzes* [Decomposition Symptoms of Wood], Berlin 1878, page 149.

[Chapter Six. Enhancement of The Forests By Care of The Shrubs And The Groundcover]

69) To page 211. "Aus dem Walde" [Out of The Forest] X 24.

70) To page 212. If somebody thinks of the non-woody plants as not being worth detailed consideration and some care, I would like to recommend for him to read the essay by Burckhardt: "Die Waldflora und ihre Wandlungen" [The Forest Flora And Its Changes] in *(Aus dem Walde V.)*. Burckhardt wants to inspire the reader "to look frequently also at the small plants in the forest, for use and benefit of the economy, **for seasoning and adornment of forest life,** as well as for worship of the great seed masters in Mother Nature."

71) To page 213. A very rich literature, from von Wangenheim and Du Roi up to James Bouth deals with the introduction of foreign tree species. Whoever is not yet very familiar with the subject and wants to undertake tests should stay with the official publications of the forstliche Versuchsanstalten [Research Institutes for Forestry] (in the *Danckelmann Journal*), and additionally consult the gardening literature concerning the aesthetic value. By the way, it is remarkable that Prince Pueckler wanted to allow foreign tree species only in the garden, not in the park (except for a few species established long ago). However, Petzold said in the already mentioned writing dedicated to the Prince's memory: "Since the establishment of the Muskau Park so many beautiful trees have been introduced and have established themselves with us, that I am convinced the Prince would certainly have adopted them if he had had them at his disposal at his time, especially the excellent tree species from North America, which are completely hardy in our climate. Just as the Prince composed his plantings out of the most heterogeneous tree and shrub species, so also nature plants its forests in North America; therefore, these have such a great appeal in respect to landscape, and they give an appeal, a glow of colors and a splendor of colors to the landscape, especially in fall, by generation of so many beautiful trees colored in all the nuances of red, orange and yellow, which we do not know at all, because we have too much green in the scenes of our landscape, but in fall we have too much yellow."

Even if I admitted that Petzold was right with his conviction, nevertheless Prince Pueckler certainly would not have let the strangers invade the forest beyond the park.

72) To page 217. Burckhardt: *Aus dem Walde* [Out of The Forest], booklet I.

73) To page 217. Count Edgar zu In und Knyphausen: "Aus dem Park in den Wald" [Out of The Park Into The Forest]. (Aus dem Walde, booklet VIII.)

74) To page 217. Nowadays so many **varieties and subvarieties of our native tree species** are distributed in garden and parks that their number cannot be overlooked any more. When he published his "Versuch einer vollstaendigen Geschichte vorzueglicher Holzarten" [Attempt of a Complete History of Excellent Tree Species] in 1783, von Burgsdorf still believed that he had to mention also the varieties. At his time he knew of one variety each of both of the oak species, and of four varieties of the beech which varied in foliage, he did not know of any which varied in stature. Nowadays, just the catalogue of the Count von Arnim's Tree Nursery in Muskau Ober Lausitz lists 39 varieties of *Quercus robur,* 15 of

Q. petraea, and 19 varieties of *Fagus sylvatica.* The names had been heavily confused by the horticultural traders in the beginning, which was clarified a lot (also in respect of the foreign tree species) by the establishment of the marvelous Arboretum in Muskau in the year 1859. Whoever is particularly interested in dendrology cannot leave the Muskau Park unnoticed. There is information on that in: Petzold and Kirchner: *Arboretum Muskaviense* [The Arboretum in Muskau], (Leipzig at W. Opetz). More recently than the publication mentioned above are writings with similar contents by Jaeger: *Die Ziergehoelze* [Decorative Trees and Shrubs], and Lauche: *Deutsche Dendrologie* [German Dendrology], (Berlin, Paul Parey).

But all of these deal as much with foreign tree species as with native ones. I do not know yet of a book which gives completely the attention they deserve to the varieties of the native tree species, but I found single bits of valuable information scattered here and there, for example by Karl Koch on the "Schlangenfichte" [Variety of *Picea excelsa*] (*Vorlesungen ueber Dendrologie* [Lectures On Dendrology], Stuttgart 1875) and by Petzold on the Pyramiden-Eiche [Quercus robur 'Pyramidalis'] in (Aus dem Walde booklet VIII).

[Final Remarks]

75) To page 226. *Forstliche Bauten und Gaerten* [Forestry Buildings and Gardens]. See my essay in the *Year Book* of the Schlesischer Forst-Verein [Silesian Forest Society] for the year 1878: "Forstaesthetische Reiseergebnisse" [Forest Aesthetic Travel Results].

76) To page 226. An excellent and richly supplied book on room decoration is available: G. Hirth: "Das deutsche Zimmer der Renaissanze, Anregungen zu haeuslicher Kunstpflege" [The German Room of The Renaissance, Ideas for Domestic Care of Art], Muenchen 1880. In this book there are also some drawings of such simple household equipment that they can be given to even a rural carpenter as models for the woodcutter's housing.

77) To page 226. Our German "Forstmanns-sprache" [Foresters' Language] owes its contemporary form to unified efforts of well-known foresters, after the necessity to examine and to revise the so-called "Forstkunstsprache" [Forestry Jargon] had been pointed out by Behlen at the VIII. Meeting of German Agriculturists and Foresters. See "Allgemeine Forst- und Jagdzeitung" [General Forestry And Hunting Journal] 1845 and the following years.

Since then, a new revision with the goal to eradicate quite a few unnecessary foreign words has been suggested several times, most recently by Borggreve.

Not long ago I found as a title of an "Original-Artikel" in the "Centralblatt" [Central Paper]: "Quiescirung der Gemeindeforstwarte und Alimentirung deren Relicten" [Pensioning of The Municipal Foresters and Maintaining of Their . . . ?]. This one line proves better than a long essay could the necessity of a renewed revision of our language. Unfortunately I find myself being guilty of having used quite a few unnecessary foreign words in this book.

78) To page 226. H. Burckhardt: "Jagd- und Waldlieder" [Hunting and Forest Songs], Hannover 1866.

79) To page 226. "Jagd-Signale und Fanfaren" [Hunting Signals and Fanfares], J. Rosner, Pless O./S.

Review of Forest Aesthetics, Third Edition, 1911.

Forest Scientific Central Paper, a publication of the Royal Bavarian Forestry Experiment Station. Edited by: Dr. Hermann von Fuerst, Royal Bavarian Forestry Director and Retired Director of the former Forestry Academy at Aschaffenburg. Berlin: Paul Parey, Publisher.

III. Literature Review Number 78
Forest Aesthetics, Third Edition, by Heinrich von Salisch. With 133 Figures. Berlin: Julius Springer, 1911. 434 pages.

At two major meetings of the German Forest Society, forest aesthetics was listed on the agenda: in the year 1905 in the city of Darmstadt, when the first proposals of the speakers von Salisch and Walther were unanimously accepted: "Management of the forests according to aesthetic aspects has to be considered as a need grown out of the economic and social conditions of modern times." Also in 1906 in Danzig, when the following sentence proposed by the Forest Management Director was accepted nearly without opposition: "It seems to be appropriate that management of forest beauty should be taught in academic lectures at forestry academies." However, the further proposal, which demanded to ask the appropriate ministries of the German states to initialize the holding of special lectures on forest aesthetic care at universities, was not as successful. Nevertheless, the speakers had to appreciate that the idea expressed and explained in detail by them was met by a sympathy hardly ever known before.

In the meantime, *Forest Aesthetics* by H. v. Salisch, published for the first time in 1885, has been published in its third edition, which would not have become necessary without the increased interest in the subject, caused by the author and others. The structure of the material has remained the same as it was for the review of the second edition (*Forstwissenschaftliches Centralblatt* [Forest Scientific Central Paper], 1902, pages 525-527). However, there has been a considerable increase in the amount of material and in considered professional aspects. Also, the number of figures has been increased.

The author does not limit his remarkable discussions to the perceptions close to him and to the experiences of his own property and his home in a narrow sense. It is understandable, however, that these take up a lot of space, and without doubt, the book would have reached a greater publicity if forest scenes and management types of other regions had been discussed in as much detail.

In order to illuminate the extraordinary richness of the subject rather than to criticize the book, I, being a southern German forester, would like to list a few things that I missed in Part I, Section B, Beauty of Nature.

In Chapter 8 on the "Aesthetic Value of Tree Species," the excellent beautiful groups and single trunks of noble hardwoods, especially Sycamore Maple [*Acer pseudoplatanus*], deserve to be mentioned which are so pleasant to the eye in meadows, at forest edges, on mountain clearings and mountain meadows in the Bavarian high mountains and in the "Bayerischer Wald" [Bavarian Forest].

In Chapter 11, the Bavarian-Bohemian forested mountains with their mighty spruces, firs, and beeches should be listed as an extraordinary rich area of beautiful trees among the "Nature Monuments from the German Tree World."

By the way, I would like to draw attention to a small error that may have been caused by a source referred to (by von Salisch). It is that the yew, called the "oldest German tree" and presented in a picture and mentioned in the text on pages 108 and 143, located in "Berggruendle" in the Allgaeu, is not growing at an elevation of 1520 m above sea level as stated in the text, but at an elevation of 1125 m (that is 3853 Bavarian feet). This can be seen in the very reliable "Vegetationsverhaeltnissen Suedbayerns" [Vegetative Conditions of Southern Bavaria] (page 519) by Otto Sendtner.

More disturbing, however, is a gap in Part II, Applied Forest Aesthetics. In Chapter 16 the "Silvicultural Systems and Their Aesthetic Effects" are discussed. The author prefers the high forest, especially in the form of shelter tree forest, over all other silvicultural systems, even the plenter (selection) forest. But he does not mention the silvicultural system which lies between plenter forest and the more or less monotonous high forest with clearcuttings, the one offering the most beautiful forest scenes, and that is the regulated (regular) femel-cutting forest, especially the group-wise femel cutting as it is practiced in Bavaria as well as other places in Southern Germany.

This type of stand regeneration—also being a form of high forest—has evolved there naturally, has been promoted by able foresters, and has been founded systematically and taught by extraordinary men of science, especially by Karl Gayer. It has worked well for a long time and, besides its forestry advantages, has produced forest scenes which can satisfy every forest aesthetician. Certainly, the very honorable author of *Forest Aesthetics* would like it, too, who confesses in Chapter 19 to be himself a friend of natural regeneration wherever conditions give certain hope for success. In our mixed forests, made up of spruces, firs, and beeches, there is only one threat to the success of group-wise or edge-wise femel cutting, and that is the necessity of too quickly liquidating stands in the regeneration process.

Mr. von Salisch proves several times for particular cases that the forest owner does not need to make large sacrifices for the forest's beauty, but that frequently measures which improve forest beauty also offer advantages for the user. When publishing the book, he may not have known yet of the reproach that recently has been made against the Hessen State Forest Administration, that is, that their forest aesthetics has more weight than efficiency and profit, and costs them productivity (*Forstwissenschaftliches Centralblatt* 1911, Part 5, Literary Reviews: "The Grand Duchy Hessian State Silviculture.")

The author of *Forest Aesthetics* otherwise also demands primary consideration of the net profit to the owner. He is not even hostile towards the land's net profit. He only demands, together with Wilbrand and others, that besides the factors creating value and increasing it, the forest's beauty and its maintenance should also be included in the calculation. Towards the end of Chapter 18 where he discusses this subject, the words are found: "Whether one is a supporter of the highest net profit theory or not, in any case, the forester might want to keep his idealism." Characteristic in this sense is also the distinction between park and forest, discussed and argued in detail in Chapter 29. A difference between the two of them exists and should be recognized from looking at them; the landscape gardener establishes and takes care of the park as well as the park-like garden, even if a forester often is consulted for the project, and even if often he is given the task of establishing "park-like landscapes" and allees.

The author mentions a number of extraordinarily beautiful allees; besides the ones he mentions, the Lichtenthal Allee near Baden-Baden and the allee between Salzburg [Austria] and Hellbrunn also deserve attention.

In an epilogue, v. Salisch summarizes briefly what he understands by the "essence of forestry art." "The most important thing is a management system in accordance with the state of knowledge of the science and art, the tending of tree and shrub species fit for local conditions, maintenance of venerable old trees, and the access of stands by paths or roads which lead into the interior. If there is an opportunity to add adornment in the form of nice buildings, by aesthetic design of meadows and open water, or by rocks and stones, it should not be wasted.

Vistas may allow the eye to roam far away."

From the last sentence of the last (39th) chapter, it can be seen how high the author values the task given to foresters and forest owners when he says: "The forester should not only care for the beauty, he also should educate forest visitors to understand and appreciate the offered forest magnificence."

However, for this a sense and understanding of the beautiful is necessary; in order to gain it and to increase it, hardly anything else is more suitable than the study of this book in its new edition, which we can recommend very warmly to the forestry audience as well as forest owners.

Raesfeldt

Index

Administrative units [compartments] 131
Adornment [decoration] 27, 63
Alder 128, 207
Alder, European 101
Alder, white 112, 266
Alder, hazel 112
Allee, fruit tree 254
Allees 246
Allees, closed 248
Allees, open 250
Allees, of different species 251
Altum 37
American linden [basswood] 111
Animals 38
Apple tree 82
Architecture 218
Arndt, E. M. 119
Art, beauty in 34
Art, cultural 1
Art, forestry 1, 172, 218
Art, landscape 1, [gardening] 218
Art [painting] 99
Arve [*Pinus cembra*] 98
Ash 79
Aspen 101, 210
Associations for Embellishment 221, 245, 295
Atmospheric perspective 51

Background 293
Back-stepping colors 41
Bank [pond or stream] planting 210
Beautiful, the purely 33
Beauty, in art 34
Beauty, increase in 178
Beauty, natural 34
Beauty, types of 33
Beech 5, 74, 152, 165, 265
Beech, blood 78, 278
Bees 202
Beekeeping 202
Beissner 92, 94
Berries 199
Berthold 309
Birch 101, 165
Boden 184
Bolle, Dr. C. 80
Borch, von der 12, 225
Border trees, removal of 224, 304
Bordering of roads 246
Borggreve 181
Bratranek 76, 113, 271
Bridges 23
Broom [shrub] 107
Building 219, 293
Burckhardt 13, 82, 160, 239, 258, 281
Burgsdorff, von 80
Butterflies 281

Care of the stand 187
Carrefour [four-cornered intersection] 133
Cherry, bird 105
Cherry, sweet 82
Chestnut 160
Clearcutting 155
Clearcutting, strip 144
Color, contrast 40, 41, 53
Color, combination of 42
Color, theory of 38, 124, 155, 160
Colors, cool and warm 41

Colors, forward and backward stepping 41
Colors, mixture of 48
Colors, pure and mixed [blended] 44
Combination culture [forest and farm crops] 181, 186
Composite forest 159
Conifers 53, 167
Contrast 40
Copse [light grove] 159, 225
Cotta 125
Creek 56
Crossroad [intersection] 132, 239

Danckelmann 158
Darwin 16
Deciduous trees 165
Decoration [adornment] 27
Deer [game] 122, 281
Denzin 6, 134
Distribution of forests 123
Domaszewski, von 56
Douglas-fir 113
Drainage ditch 205, 233
Dwarf pine [*Pinus pumilio*] 109

Ebermeyer 196
Eilenriede 258
Elder, black [*Sambucus*] 106
Elm 81
English ivy 109, 282
Entrenchments [earthworks] 290
European silver fir 97
Evening light [twilight] 52
Euonymus europaeus 106
Experimental forest 269
Eye 38

Fechner 23, 147
Fences 212
Fickert 261
Field 128, 181, 212
Fir, European silver 97
Firebreak 134
Fischbach, von 15
Fish 211
Flowers 152, 283
Footpath 237
Foreground 293, 297
Foreign tree species 110, 267
Forest gardener 88
Forest, high 152, 303
Forest litter utilization 199
Forest, low 160
Forest, luxury 220
Forest margin 208
Forest pasture 202
Forest, primeval 149
Fragrance of the forest 99, 113
Freedom of nature 46, 121
Friedrich Wilhelm IV 229, 261, 288
Fruit tree 82, 305

Garden 271
Gayer 13
Geology 54
Germination beds 185
Gilpin 12, 68, 111, 140, 164, 299
Gleam [sheen] 44
Goethe 2, 20, 62, 221, 290
Golden section 25
Good, the 21, 31
Gracefulness 33
Grave markers, megalithic 291
Grazing 199
Grazing forest 150
Groeditzburg [castle] 167
Ground cover 160, 283
Ground moraine [till] 54
Group of trees, small [copse] 218
Groups of natural regeneration 301
Grunewald 150, 215, 282
Guse 5
Guttenberg, von 15, 142, 152

Hagen, von 179
Hampel 183, 269
Hardwoods [deciduous] 165

Hardwoods, soft 99, 160
Hartig, Georg Ludwig 70, 72, 181
Hartig, Robert 178, 260, 274
Hartig, Theodor 18
Hawthorne 106
Hazelnut 106
Heath [family] 108
Heath landscape 124
Hedgerow 126, 234
Hedges 213
Hedges (Pueckler hedges) 127, 235, 246
Hegel 34
Hempel 15
Hermann 34, 36, 45
Hertha Beech 261
Heyer 5, 93
High forest 152, 303
Highest net revenue, teaching of 172
Horsechestnut 111
Hunt, the 202
Humboldt 30
Humor 33

Idea associations 31
Inscription 242
Interest rate 177
Irrigated meadows 205

Jaeger (Court gardener) 77
Johanna's Height [tower] 222, 294
Judeich 172
Juniper 107

Kaiser Wilhelm I 72
Kessler 89, 122
King Friedrich Wilhelm IV 229, 261, 288
King Wilhelm I 72
Klipstein Oak 264
Klopstock 115
Koenig 13, 257, 259, 266
Kraft 15, 215
Krause 15
Kratzkau 229, 231
Lakes 131
Landscape, park-like 225
Landscape, winter 53, 160
Larch 97, 184
Leaf fall 196
Lenne 229, 302
Leuer 223
Linden 99, 167
Linden, American [Basswood] 111
Livestock 201
Locust, black 111
Looping paths [round trip roads] 237
Lopping system 162
Low forest 160
Luetetsburg 274
Luxury forest 220

Maintenance of old trees 258
Management plan 174
Maple 80
Masius 13, 97
Meadows 128, 202, 246
Middleground 293
Mixed stand 168
Moltke, von 10, 56, 116, 231, 283, 286
Monument 290
Mountain pastures 202
Mountainash 105
Mountains 121, 292
Mudslide 58
Mushrooms 199
Muskau 214, 291

Naming trees 267
Native species, variety of 276
Neumeister 131, 172, 243
Ney 191
Notebook 267
Nursery 185

Oak 66, 151, 198
Oak, pyramid 74
Oak, red 73, 111
Oerstedt 19, 64
Ownership, pride of 10

Paradise 147, 154
Park 121, 213
Park-like landscape 225
Park-like managed forest 214
Parliament, Upper Chamber of [Prussian] 15, 215
Pastures, mountain 202
Path [or road] 142, 237
Paths, looping [round trip roads] 237
Pear tree 82, 228
Peltz, art teacher 301
Perception 38
Petzold 74, 258, 291
Pfeil 7, 176, 246
Picturesque 33, 68
Pine 52, 82, 155
Pine, pitch 112
Pine, Siberian stone 98
Pine, Weymouth [eastern white] 112, 116
Plantations 13, 180
Planting 180
Planting, mound 185
Planting pattern 181
Pollarding 162
Ponds 207, 211
Poepel 177
Poplars 100, 116
Postel thinning method 189, 302
Pot-pruning 254
Pressler 172, 173
Pride of forest ownership 10
Primeval forest 149
Private forest owner 10
Pruning 193, 200, 231
Pueckler 13, 108, 155, 158, 183, 212, 214, 217, 227, 275, 292
Pueckler hedge 127, 235, 246
Pure stand 168
Pyramid oak 74, 277
Pyramid poplars 162
Pyramid trees 277, 303

Quaet-faslem 151

Ratzeburg 88
Red oak 73, 111
Regeneration 179
Regeneration, groups of natural 296
Regular division 131
Removal of border trees [strip cutting] 224, 304
Riehl 120
Road [or path] 142, 237
Road, round trip 237
Road 131
Roedler 8
Rose 106
Rossmaessler 13, 16, 85
Rotation 6, 170
Row thinning 186
Rueckert 90, 103
Ruins 290

Scheppe-Allee 89
Schiemann 305
Schiessbeere [cathartic shrub] 106
Schiller 34, 56
Schleiden 13, 116, 121, 152
Schneider, Dr. K. E. 2
Schwarz, G. 322
Seasons 54
Secondary roads 237
Secondary uses 196
Seeding [sowing] 180
Seed-tree cutting 156, 159, 293
Selection [Borggreve] thinning 191
Selection [plenter] forest 151, 304
Selenka 28, 63
Serpentine [switchback] roads 142
Shade intolerant species 166
Shade tolerant species 166
Shelter plantings [shelterbelts] 126
Shelter-selection method 166
Shrubs 105, 279
Signs, trail 242
Silvicultural systems 149
Small geometric cuttings 141
Soil use 119

Species 63
Species, foreign 110, 267
Species, varieties of native 64, 267
Spruce 94, 153, 181
Stand border 247
Stand, care of 187
Stand, mixed 168
Stand, pure 168
Stone blocks 284
Stones 54
Strip clearcutting 144
Stump removal 186
Style 20
Sublime 33

Tanbark forest 161
Tegel 80, 113
Telegraph wires 256
Thiergarten [Berlin Zoo] 98, 241, 257, 303
Thinning 187
Thinning, Postel method 189, 302
Timber cutting 112, 143, 291
Tower, lookout 295
Tragic, the 33
Truth 20
Typical 21

Ugly, the 33
Understory 166, 180, 280
Uniform 44
Unity in variety 25
Upper Chamber of [Prussian] Parliament 15, 215
Usefulness 23

Viburnum 106
Vines 109
Virgil 66
Virgin [primeval] forest 150
Vischer 3, 75
Vista 155, 159, 294
Voices of the forest 113

Walnut tree 111
Wang 57
Wassdorff 301
Waterfalls 287
Water surface 130
Wavy lines 27
Weise 214
Weymouth [eastern white] pine 112, 116
Wilamowitz-Moellendorf, Count von 274
Wilbrand 14, 166, 173, 219, 168, 299
Wildungen, von 170, 242
Willow 103, 164
Wind 116
Windfall wood [for firewood] 199
Winter landscape 53, 160

Yew 98, 241, 266

Zoo [wildlife preserve] 282
Zoo, Berlin [Thiergarten] 100, 241, 244, 257, 302
Zoo, Nesigode 105, 130, 301, 302